SPATIAL ORGANIZATION IN EUKARYOTIC MICROBES

Special Publications of the Society for General Microbiology

Publications Officer: Dr Duncan E.S.Stewart-Tull, Harvest House, 62 London Road, Reading RG1 5AS, UK

Publisher: Academic Press

1. Coryneform Bacteria
2. Adhesion of Micro-organisms to Surfaces
3. Microbial Polysaccharides and Polysaccharases
4. The Aerobic Endospore-forming Bacteria: Classification and Identification
5. Mixed Culture Fermentations
6. Bioactive Microbial Products: Search and Discovery
7. Sediment Microbiology
8. Sourcebook of Experiments for the Teaching of Microbiology
9. Microbial Diseases of Fish
10. Bioactive Microbial Products 2: Development and Production
11. Aspects of Microbial Metabolism and Ecology
12. Vectors in Virus Biology
13. The Virulence of Escherichia coli
14. Microbial Gas Metabolism
15. Computer-Assisted Bacterial Systematics
16. Bacteria in Their Natural Environments
17. Microbes in Extreme Environments
18. Bioactive Microbial Products 3: Downstream Processing

Publisher: IRL Press

19. Antigenic Variation in Infectious Diseases
20. Nitrification
21. Carbon Substrates in Biotechnology
22. Gene Structure in Eukaryotic Microbes
23. Spatial Organization in Eukaryotic Microbes

This book is based on a symposium of the Cell Biology Group of the SGM held in December 1986.

SPECIAL PUBLICATIONS OF THE SOCIETY FOR GENERAL MICROBIOLOGY
VOLUME 23

SPATIAL ORGANIZATION IN EUKARYOTIC MICROBES

Edited by

R.K.Poole

King's College London, University of London,
Kensington Campus, Campden Hill Road,
London W8 7AH, UK

A.P.J.Trinci

School of Biological Sciences, Williamson Building
University of Manchester, Oxford Road,
Manchester M13 9PL, UK

1987
Published for the
Society for General Microbiology
by

IRL PRESS
OXFORD · WASHINGTON DC

IRL Press Limited
PO Box 1
Eynsham
Oxford OX8 1JJ
England

British Library Cataloguing in Publication Data

Spatial organization in eukaryotic microbes.
—(Special publications of the Society for
General Microbiology, ISSN 0197–1751; v. 23)
1. Micro-organisms 2. Eukaryotic cells
I. Poole,Robert K. II. Trinci,A.P.J.
III. Society for General Microbiology
IV. Series
576 QR1

ISBN 1-85221-053-2 (hardbound)
ISBN 1-85221-052-4 (softbound)

Cover illustration. The cover design is based on Figures 4 and 6 of Chapter 1 and shows a simulation cytoskeleton at the apex of a regenerating *Acetabularia* plant and a periodic calcium pattern.

Printed by Information Printing Ltd, Oxford, England

Contents

Abbreviations

cdc	cell division cycle
DAPI	4′,6-diamidino-2-phenylindole
DIF	differentiation inducing factor
cpDNA	chloroplast DNA
mtDNA	mitochondrial DNA
HMM	heavy meromyosin
MF	microfilaments
NETO	new end take off
RIM	reflexion interference microscopy
TLCCD	thin-layer countercurrent distribution
ts	temperature sensitive

Preface

During the last two decades a great deal has been learnt about the genetics, biochemistry and molecular biology of the regulation of temporal events in microbial cells and in particular about the regulation of the cell cycles of prokaryotic and eukaryotic microorganisms. By contrast, comparatively little is known about the mechanisms involved in regulating the spatial organization of cells. The Cell Biology Group of the Society of General Microbiology therefore decided to hold a meeting in Swansea on 'Spatial Organization in Eukaryotic Microbes', both to summarize our present knowledge of this subject and to stimulate further research into this important aspect of cell biology. The present volume is based upon the proceedings of this meeting.

The importance of cytoskeletal elements is evident in a number of chapters. In the first, Brian Goodwin discusses how the spatial organization of cells may be regulated and assesses the role of Ca^{2+} in modulating the mechanical state of the cytoskeleton of *Acetabularia* by influencing the polymerization of microtubules and actin. Jeremy Hyams describes changes in the location of microtubules and actin during the cell cycle of *Schizosaccharomyces pombe*, and draws our attention to the correlations between actin location and septation, and between actin location and cell elongation. William Birky details the importance of the cytoskeleton in partitioning organelles between daughter cells during cell division and describes the role of microtubules in organelle migration.

Conrad King reports the role of cytoskeletal elements in the crawling polarized movement of amoebae, whereby the cytoskeleton is locked onto the substrate to provide the necessary grip. Paul Markham examines the evidence that Woronin bodies may be involved in the spatial organization of fungal hyphae, but finds little to support the hypothesis that these structures are involved in regulating hyphal polarity. Their ability to move into and block septal pores is well-established but whether a contractile system or some other mechanism operates to effect such movement remains an open question.

The use of vibrating probes to detect electrical current around the tips of fungal hyphae is described by Franklin Harold who shows that the appearance of new zones of inwardly directed current precede and predict the appearance of cell polarity and branching in *Achyla*. Polarity in this system can be explained in terms of the non-uniform distribution along hyphae of proton-translocating ATPases and amino acid uptake systems. Neil Gow supports the general hypothesis by demonstrating that the imposition of an electrical field around fungal hyphae influences their direction of growth, and he suggests that this may result from the relocation of charged proteins in the membranes. Cell polarity is readdressed by Donald Watts, who emphasises the role of morphogens in providing undifferentiated cells with positional information.

The Group and the Editors take this opportunity to thank all who spoke and attended, and the meetings staff of the Society for their excellent support.

We hope that this selection of topics illustrates the thrust of research in this challenging field and that the well-referenced chapters will provide a starting point for those who wish to read more widely.

R.K.Poole
A.P.J.Trinci

Contributors

C.W.Birky,Jr
Department of Molecular Genetics, The Ohio State University, 484 West 12th Avenue, Columbus, OH 43210, USA

C.Brière
Laboratoire de Biologie Quantitative de la Morphogénèse Végétale, Ecole National Supérieure Agronomique, 145 Avenue de Muret, F-31076 Toulouse, France

J.H.Caldwell
Department of Molecular and Cellular Biology, National Jewish Center for Immunology and Respiratory Medicine, Denver, CO 80206, USA

A.J.Collinge
Microbiology Department, King's College, Campden Hill Road, London W8 7AH, UK

B.C.Goodwin
Developmental Dynamics Research Group, Department of Biology, The Open University, Milton Keynes, MK7 6AA, UK

N.A.R.Gow
Department of Genetics and Microbiology, University of Aberdeen, Marischal College, Aberdeen AB9 1AS, UK

I.M.Hagan
Department of Botany and Microbiology, University College London, Gower Street, London WC1E 6BT, UK

F.M.Harold
Department of Molecular and Cellular Biology, National Jewish Center for Immunology and Respiratory Medicine, Denver, CO 80206, USA

J.B.Head
Microbiology Department, King's College, Campden Hill Road, London W8 7AH, UK

J.S.Hyams
Department of Botany and Microbiology, University College London, Gower Street, London WC1E 6BT, UK

C.A.King
Department of Zoology and Cell Biology, University College London, Gower Street, London WC1E 6BT, UK

S.K.Maciver
Department of Zoology and Cell Biology, University College London, Gower Street, London WC1E 6BT, UK

P.Markham
Microbiology Department, King's College, Campden Hill Road, London W8 7AH, UK

J.Marks
Department of Botany and Microbiology, University College London, Gower Street, London WC1E 6BT, UK

P.S.O'Shea
Developmental Dynamics Research Group, Department of Biology, The Open University, Milton Keynes, MK7 6AA, UK

R.K.Poole
Microbiology Department, King's College, Campden Hill Road, London W8 7AH, UK

W.J.A.Schreurs
Department of Molecular and Cellular Biology, National Jewish Center for Immunology and Respiratory Medicine, Denver, CO 80206, USA

D.J.Watts
Biochemistry Department, University of Sheffield, Sheffield S10 2TN, UK

CHAPTER 1

Mechanisms underlying the formation of spatial structure in cells

B.C.GOODWIN, C.BRIERE[1] and P.S.O'SHEA

Developmental Dynamics Research Group, Department of Biology, The Open University, Milton Keynes, MK7 6AA, UK and [1]Laboratoire de Biologie Quantitative de la Morphogénèse Végétale, Ecole National Supérieure Agronomique, 145 Avenue de Muret, F-31076 Toulouse, France

---

Introduction

Growth and morphogenesis in higher plants result from a complex interplay of processes, chief among which are rate and orientation of cell divisions and specific, controlled changes of shape of individual cells. These require spatial control over intra- and inter-cellular events affecting the mechanical state of the cytoskeleton and the cell wall. We have some understanding of how simple changes in cell shape may be regulated by, for example, hormone-induced changes in microfibril orientation which alter the ratio of lateral to longitudinal cell expansion (Ridge, 1973, 1975); but the origin of more complex cell shapes as in root hair cells or stellate cells is poorly understood. Localized weakening of cell walls must presumably arise and localized expansion follow; for root hair cells this is a process of tip growth. Investigating these processes in multicellular systems is, however, extremely difficult.

We have therefore used the unicellular marine alga, *Acetabularia mediterranea,* to study the control mechanisms governing the detailed and regular changes of shape which this cell undergoes during the regeneration of apical structures, mediated via the cytoskeleton and the plasmalemma. What particularly recommends this species for morphogenetic studies is its structural simplicity, the spatial symmetry of the generated forms, and the ease with which it lends itself to physiological and morphological investigation. There is also now a very extensive literature on many aspects of the growth and regeneration of this species (Bonnotto *et al.*, 1979; Woodcock, 1977). Furthermore, present evidence (Green *et al.*, 1971; Métraux and Taiz, 1977) suggests that the processes controlling growth in algal cells do not differ fundamentally from those in higher plant cells.

Research on *Acetabularia* and selected systems indicates that four important variables are immediately implicated in localized tip growth, namely electrical currents, mechanical strain activation of plasmalemma ion pumps, proton or enzyme secretion and calcium localization. This evidence comes from studies of different plant species,

Figure 1. The mature form of *Acetabularia mediterranea.*

but makes a coherent picture. The assocation of electrical currents with tip growth comes primarily from the work of Jaffe (1981) and his co-workers who, by use of the vibrating probe (Jaffe and Nuccitelli, 1974), have shown in the germinating lily pollen grain and

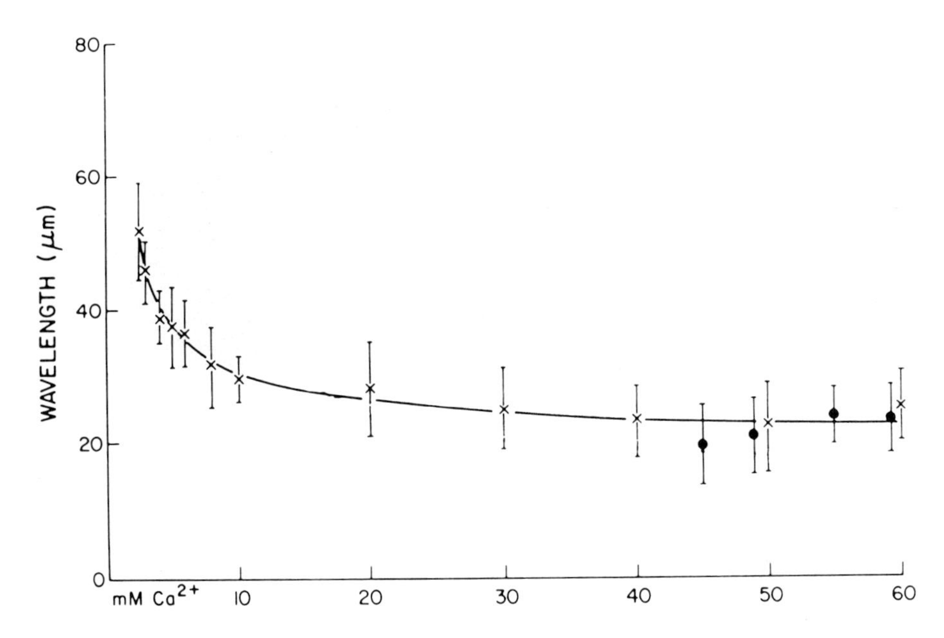

Figure 2. Changes in the wavelength of the whorl pattern as a function of external calcium in regenerating *Acetabularia* plants (×). The mean wavelengths between hairs in whorls with gaps are also shown (•).

in *Fucus* and *Pelvetia* eggs that the growth point is predicted by the point of greatest density of a steady inward-directed current that starts well before any visible growth. These and other studies (Weisenseel and Kicherer, 1981) have indicated that one of the ions carrying the current is calcium, and that there may be differential localization of calcium ion channels and pumps within the plasmalemma, creating the sources and sinks for an intracellular electric current. The electric circuit is completed by extracellular current flow, which can be measured by the vibrating probe.

Identification of the mechanical component of the growth process comes from the recognition that turgor pressure is a prerequisite for cell expansion (Green *et al.*, 1971; Ray *et al.*, 1972). There is evidence that the maintenance of turgor pressure is dependent upon pressure-sensitive ion pumps in the plasmalemma by a mechanism such as that proposed by Zimmerman and Steudle (1978). A relationship between acid-secretion, increase in wall extensibility, and cell expansion has also been suggested (Tepfer and Cleland, 1979), while there is evidence of a direct correlation between external acidification and localized cell expansion in *Chara* and *Nitella* (Lucas and Smith, 1973; Taiz *et al.*, 1981). These observations suggest, but do not yet clearly establish, a causal relationship between local mechanical strain and proton pump activation. It remains a possibility that other mechanisms of wall loosening may be involved in the process, such as hydrolase secretion. These are questions we are currently investigating in *Acetabularia* (*Figure 1*).

The occurrence of increased levels of calcium at the tip of the growing lily pollen tube has been demonstrated by Reiss and Herth (1979) using chlorotetracycline. This correlates well with the electrical current observations of Jaffe (1981) and the evidence

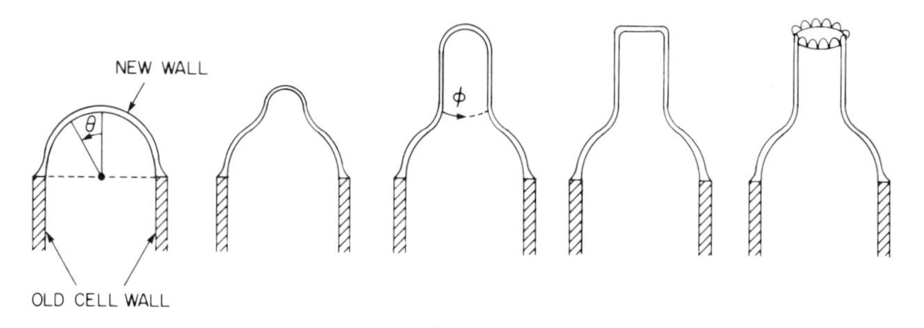

Figure 3. Sequence of morphogenetic shape change at the regenerating tip of *Acetabularia*, from initial repair after cap removal to the beginning of whorl formation.

that calcium is the major ion involved. Furthermore, Goodwin *et al.* (1983) have presented evidence that transitions between the different phases of regeneration in *Acetabularia* can be controlled by extracellular calcium concentration localized in the tip region. Experimental evidence implicating bound calcium as a 'morphogen' in whorl formation has also been presented (Harrison and Hillier, 1985).

The results of a similar study of variation in the wavelength of the whorl pattern as a function of extracellular calcium are shown in *Figure 2*. The wavelength is measured as the mean distance between hair primordia soon after they become visible in whorl formation (see *Figure 3e*). This is determined by measuring the diameter of the whorl, counting the number of hairs and calculating the mean radial distance between hairs. The wavelength decreases in a systematic manner as the external calcium concentration varies over the range in which whorl formation is possible (2–60 mM Ca^{2+}; see Goodwin *et al.*, 1983). Interesting evidence that the whorl pattern arises as a global periodicity in some variable(s) comes from the observation of incomplete whorls that form occasionally towards the upper end of the permissive calcium range. Despite the presence of extensive gaps in the pattern where hairs failed to form, the regions of continuous hair primordia had the same wavelength as that of complete whorls. This insensitivity of wavelength to deletions of pattern suggests that whorls are initiated by complete rings of periodicity in variables which could include cytosolic free calcium as a primary morphogenetic agent. The gaps may then be due to regions of the pattern where the amplitude of the periodicity falls below a threshold value for hair formation, as might be expected near a threshold of whorl formation.

These observations all make a *prima facie* case for a connection between electrical, mechanical and ionic variables, primarily calcium and protons, in the control of tip growth and morphogenesis. The experimental system which we have been studying is the sequence of regenerative events following cap removal in *A. mediterranea*. After the wound reaction and recovery, the apical cytoplasm and plasmalemma assume a hemispherical form, with a thin cell wall forming outside the plasmalemma. The first phase of regeneration is the formation of a conical tip which elongates. The geometry of these shape changes are shown in *Figure 3*, where only the cell wall is drawn. After a period of elongation, the tip flattens and a ring of hair primordia forms, as shown in the last two drawings of *Figure 3*. The hairs of the whorl, as the circular structure is called, grow and branch, producing a delicate structure which later dies, the hairs falling off the stalk and leaving a scar. From the centre of a whorl, a new tip arises and axial growth resumes. Depending upon growth conditions (ion concentrations, nutrients, light, etc) this process may be repeated a number of times, giving a sequence of whorls. The final stage of regeneration occurs when the tip flattens and a cap primordium is formed, a structure which is considerably more complex than a whorl of hairs but has the same radial symmetry. This grows radially to produce the adult cap, an intricately-sculpted form which reaches a characteristic size. Upon cessation of cap growth, the plant reaches a state of morphogenetic equilibrium (for a comprehensive description of these processes, see Puiseux-Dao, 1970).

The morphogenetic field

The question we address in this chapter is the extent to which these morphogenetic changes may be described by a field theory based upon the spatio-temporal properties of the cytoskeleton in the cortical cytoplasm of *Acetabularia*. Field equations describing the state of this structure and the kinetics of cytosolic free calcium, with which it interacts, were derived in a paper by Goodwin and Trainor (1985). These equations are basically the same as those first used by Odell *et al.* (1981) to describe certain morphogenetic processes in animal embryos, with modifications arising from our interpretation of cytoskeletal and calcium dynamics. The equations are applicable to any type of cell, and so in principle could describe the basis of morphogenesis in any organism, with secondary modifications to take account of features characteristic of different types, such as plants or animals, unicellular or multicellular. Applications of this analysis to a variety of different morphogenetic processes have been described by Oster *et al.* (1983) and Murray and Oster (1984). The proposition, then, is that mechanochemical fields based upon cytoskeletal-ionic interactions may be the universal basis of morphogenesis.

An implication of this proposition is that fields based upon the kinetics of diffusion-reaction may not play the primary role in morphogenesis proposed first by Turing (1952), so that there may be no 'morphogens' in this sense. This would account for the remarkable difficulty that has been encountered in identifying such substances in developing organisms.

Starting with the equation derived by Goodwin and Trainor (1985), we carried out a study of their behaviour using a modified finite-element method to describe the inter-

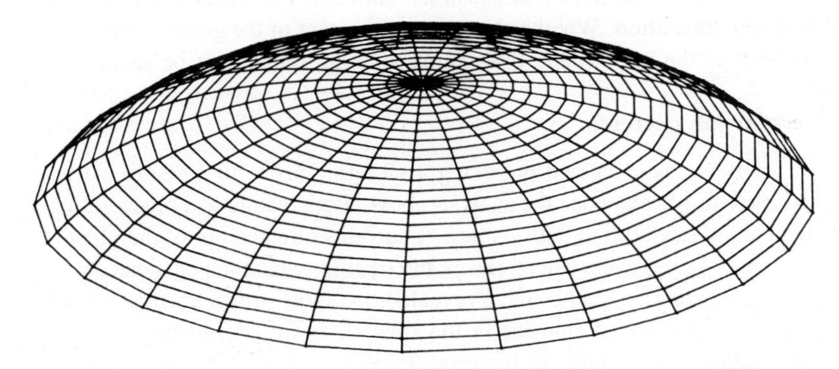

Figure 4. Mesh of viscoelastic filaments simulating the cytoskeleton in the cortical cytoplasm at the apex of a regenerating *Acetabularia* plant.

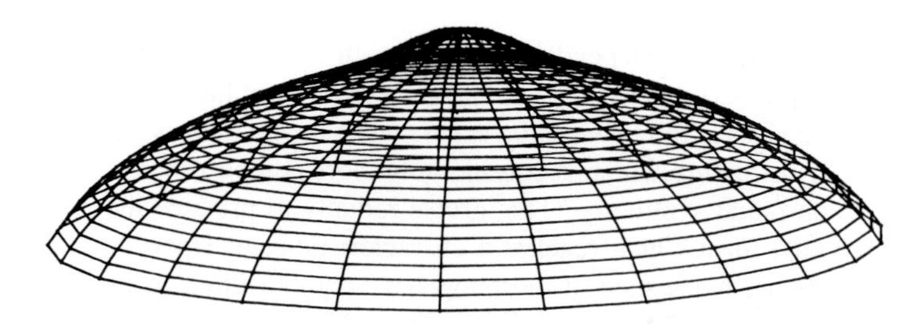

Figure 5. Initiation of tip formation by a bifurcation in the cytoskeleton-calcium system, with a decrease in the elastic modulus of the wall at the apex and buckling out due to turgor pressure.

active dynamics of the cytoskeleton and calcium in the cortical cytoplasm (the thin gel-like, non-streaming region of cytoplasm lying immediately beneath the plasma membrane, which is pressed against the cell wall by turgor pressure from the central vacuole). We began with an analysis of the behaviour of a two-dimensional shell, simulating the regenerating apical tip of an *Acetabularia* plant. This shell is described by a network of finite elements as shown in *Figure 4*, each of which is assumed to obey the visco-elastic and calcium equations of Goodwin and Trainor (1985). For purposes of computation, these take the form described elsewhere (Goodwin and Brière, 1987). The cell wall, external to this network, was described simply by a substance with a variable elastic modulus whose value was defined by a function of the mechanical strain in the underlying cortical cytoplasm. The reasoning behind this coupling is explained in Goodwin and Trainor (1985): where the cortical cytoplasm is strained, so will be the plasmalemma, which could then either activate proton pumps, resulting in local wall acidification and softening; or facilitate hydrolase secretion. The result of the simulation is shown in *Figure 5*. The geometry together with the boundary conditions have the consequence that maximum strain occurs at the tip, the cell wall elasticity decreases here and bulges out under the action of the outward-acting turgor pressure.

However, we now face the question of how the mechanochemical model produces the whorl pattern. We have shown in a simulation of a one-dimensional version of the non-linear equations that growth of a periodic mode does indeed occur for parameter values satisfying the bifurcation condition, as shown in *Figure 6*. However, we have not yet demonstrated this in two dimensions, although we expect to obtain this in the full non-linear simulation. Whether or not the geometry of the growing tip dictates the whorl pattern as the most stable bifurcating solution remains to be seen.

Extracellular ion currents

A recent study we have carried out using a vibrating probe to measure extracellular electrical currents has extended the analysis above. Both mature and regenerating plants in the light show large (up to 380 μA.cm^{-2}) currents directed out of the plant (coventional current polarity) everywhere except the rhizoid, where the currents are directed inwards. Ion substitution studies have revealed that most of this current actually consists of an outward flux of chloride at the rhizoid, inwards elsewhere. This current is greatly reduced in the dark, to between 0 and 4 μA.cm^{-2}. Previous experimental studies (Goodwin *et al.* 1983) had shown that regeneration can be selectively inhibited

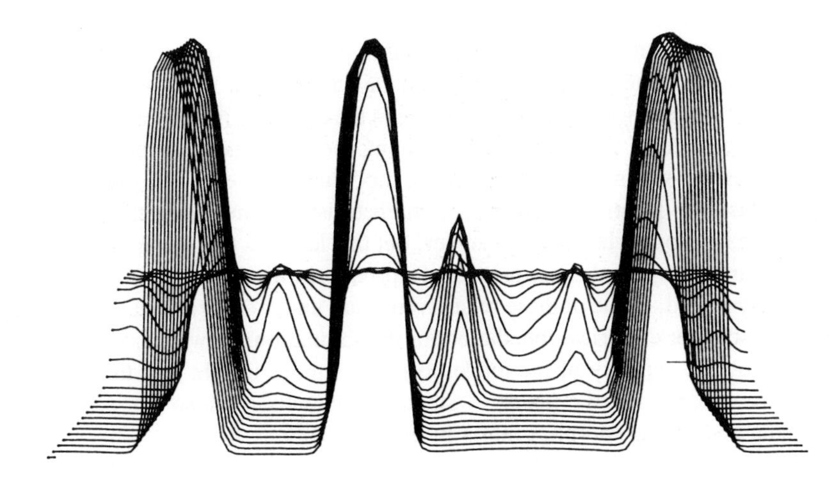

Figure 6. Development of a periodic pattern in calcium in a one-dimensional simulation of the cytogel with parameter values in the bifuraction range.

at different stages (tip, whorl, cap formation) by either reduction in external calcium to specific levels in the external medium, or addition of ions such as cobalt or lanthanum which block calcium entry into the cell. It was therefore of interest to see if the addition of cobalt to the medium in which the plants were regenerating had an effect on the current.

Plants regenerate in the dark as rapidly as in the light, providing they are on a light–dark regime. Since the dark current is much smaller than the light, the effect of added cobalt (0.5 μM $CoCl_2$) was measured and found to reduce the current density by 5–15%, amounting to less than 0.4 $\mu A.cm^{-2}$, which is nevertheless a substantial calcium flux in absolute terms. A similar current reduction in response to added cobalt was observed in the light, but it was more difficult to measure accurately because of the large current levels under these conditions. A calcium outflow from the stalk and into the rhizoid thus appears to be a component of the current flux in *Acetabularia*.

A further observation on regenerating plants was very interesting. It was found that the extreme apex of the regenerating tip is electrically silent, with no net current entering or leaving the plant in this region. This is where most of the growth occurs, new membrane and cell wall being actively produced as the cell elongates. It thus seems most likely that electrical neutrality is maintained by a balanced rate of ion movement across this growth domain, with a net calcium influx here balanced by other cation efflux. However, a resolution of these details requires further study. This work will be reported *in extenso* in a forthcoming paper by O'Shea *et al.* (1987).

Thus, our combined experimental and theoretical study of a rather basic example of tip growth and morphogenesis in the unicellular giant marine alga, *Acetabularia*, provides a model system for the study of more complex forms of apical growth. There are still many questions that require resolution in the unicellular case, foremost among which are the conditions for sustained growth after tip formation in the model (*Figure 4*) and the realization of the whorl pattern, a three-dimensional version of the bifurcation in *Figure 6*. On the experimental side, major questions concern the exact nature of the ion fluxes associated with regeneration at the apical tip, the intracellular distribution of cytosolic free calcium in the tip region, and the nature of the wall softening process.

A number of interesting questions have arisen in the course of modelling these morphogenetic transformations in *Acetabularia*. Among these are the coupling between the state of the cortical strain field and the elastic modulus of the cell wall, and the transition from elastic to plastic deformations, characteristic of morphogenesis. We have handled these in a somewhat *ad hoc* manner, but this leads us to a better understanding of the subtle and unexpected consequences of a process in which dynamics and geometry are intimately linked. Some of these relationships are under experimental investigation to provide a sounder empirical foundation for the simulation. However, we are greatly encouraged by the capacity of the model to give us more insight into the nature and dynamics of morphogenetic fields in single plant cells, and so to suggest basic mechanisms underlying morphogenesis in higher plants.

Conclusion

There is evidence that plant morphogenesis involves mechanical forces originating in the cytoskeleton, strain activation of plasmalemma pumps and the transmembrane ion fluxes among which calcium and protons appears to be of primary importance. The giant unicellular marine alga, *Acetabularia*, lends itself to the study of these processes as they occur in single cells, and the experimental observations obtained with this organism point to calcium as a major variable in the control of regeneration and morphogenesis. A theoretical model of calcium-cytoskeletal interactions in the cortical cytoplasm can be modelled by a modified finite-element method to simulate the regenerative process. When cell wall elasticity is coupled to the strain field of the cytoplasm by a functional relationship representing plasmalemma pump activation and wall softening, the regenerating form of the tip is produced. A bifurcation in the mechanochemical equations can be demonstrated, which could underlie whorl formation. The model gives considerable insight into the nature of the morphogenetic field in plants.

References

Bonotto,S., Kefeli,V. and Puiseux-Dao,S. (eds) (1979) *Developmental Biology of Acetabularia*, Elsevier/North Holland Biomedical Press, Amsterdam.

Goodwin,B.C. and Brière,C. (1987) The concept of the morphogenetic field in plants. *Séminaire de Biologie Théorique Solignac (France)*, in press.

Goodwin,B.C. and Pateromichelakis,S. (1970) The role of electrical fields, ions, and the cortex in the morphogenesis of *Acetabularia mediterranea. Planta,* **145**, 427–435.

Goodwin,B.C. and Trainor,L.E.H. (1985) Tip and whorl morphogenesis in *Acetabularia* by calcium-regulated strain fields. *Journal of Theoretical Biology,* **117**, 79–106.

Goodwin,B.C., Skelton,J.L. and Kirk-Bell,S.M. (1983) Control of regeneration and morphogenesis by divalent cations in *Acetabularia mediterranea. Planta,* **157**, 1–7.

Green,P.B., Erickson,R.O. and Buggy,J. (1971) Metabolic and physical control of cell elongation rate. *In vitro* studies in *Nitella. Plant Physiology,* **47**, 423–430.

Harrison,L.G. and Hillier,N.A. (1985) Quantitative control of *Acetabularia* morphogenesis by extracellular calcium: a test of kinetic theory. *Journal of Theoretical Biology,* **114**, 177–192.

Jaffe,L.F. (1981) The role of ionic currents in establishing developmental pattern. *Philosophical Transactions of the Royal Society B,* **295**, 553–566.

Jaffe,L.E. and Nuccitelli,R. (1974) An ultrasensitive vibrating electrode for measuring steady extracellular currents. *Journal of Cell Biology,* **63**, 611–628.

Lucas,W.J. and Smith,F.A. (1973) The formation of alkaline and acid regions at the surface of *Chara corallina* cells. *Journal of Experimental Botany,* **24**, 15–28.
Métraux,J.P. and Taiz,L. (1977) Cell wall extension in *Nitella* as influenced by acid and ions. *Proceedings of the National Academy of Science, USA,* **74**, 1565–1569.
Murray,J.D. and Oster,G. (1984) Generation of biological pattern and form. *IMA Journal of Mathematics in Medicine and Biology,* **1**, 51–75.
Odell,G., Oster,G.F., Burnside,B. and Alberch,P. (1981) The mechanical basis of morphogenesis. *Developmental Biology,* **85**, 446–462.
Oster,G.F., Murray,J.D. and Harris,A. (1983) Mechanical aspects of mesenchymal morphogenesis. *Journal of Embryology and Experimental Morphology,* **78**, 83–125.
O'Shea,P.S., Goodwin,B.C. and Ridge,I. (1987) Patterns of electrical currents in normal and regenerating *Acetabularia. Planta,* in press.
Puiseux-Dao,S. (1970) *Acetabularia and Cell Biology,* Logos Press.
Ray,P.M., Green,P.B. and Cleland,R.E. (1972) The role of turgor in plant cell growth. *Nature,* **239**, 163–164.
Reiss,H.D. and Herth,W. (1979) Calcium gradients in tip growing plant cells visualised by chlorotetracycline fluorescence. *Planta,* **146**, 615–621.
Ridge,I. (1973) The control of cell shape and rate of cell expansion by ethylene and indole-3-acetic acid in shoots of *Pisum sativum. Acta Botanica. Neerlandica,* **22**, 144–158.
Ridge,I. (1975) Effects of ethylene and IAA on cell wall structure in tissue swellings. *Annals of Applied Biology,* **81**, 98–101.
Taiz,L., Métraux,J.P. and Richmond,P.A. (1981) Control of cell expansion in the *Nitella* internode. *Cell Biology Monographs,* **8**, 231–264.
Tepfer,M. and Cleland,R.E. (1979) A comparison of acid-induced cell wall loosening in *Valonia ventricosa* and in oat coleoptiles. *Plant Physiology,* **63**, 898–902.
Turing,A.M. (1952) The chemical basis of morphogenesis. *Philosophical Transactions of the Royal Society B,* **237**, 37–72.
Weisenseel,M.H. and Kicherer,R.M. (1981) Ionic currents as control mechanism in cytomorphogenesis. *Cell Biology Monographs,* **81**, 379–399.
Woodcock,C.L.F. (ed.) (1977) *Progress in Acetabularia Research.* Academic Press, New York.
Zimmermann,U. and Steudle,E. (1978) Physical aspects of water relations of plant cells. *Advances in Botanical Research,* **6**, 45–117.

CHAPTER 2

Endogenous electric currents and polarized growth of fungal hyphae

FRANKLIN M.HAROLD, JOHN H.CALDWELL and WILHELMUS J.A.SCHREURS

Department of Molecular and Cellular Biology, National Jewish Center for Immunology and Respiratory Medicine, Denver, CO 80206, USA

Introduction

Eukaryotic cells and organisms commonly drive electric currents through themselves. This somewhat mystifying phenomenon was first explored by Elmer Lund, who devoted to it much of his life's work (Lund, 1947). Lund believed that endogenous electric currents are part of the mechanism by which growing and developing organisms impose polarity upon themselves, and this hypothesis continues to provide a major impetus for research in this field.

Endogenous electric currents became experimentally accessible when L.F.Jaffe and R.Nuccitelli constructed an ultrasensitive instrument that allowed measurement of the minute electric currents generated by single cells (Jaffe and Nuccitelli, 1974). The sensing element of their 'vibrating probe' is a micro-electrode that vibrates over a span of 30 μm at some 250 cycles per second, and measures the potential difference between the ends of its excursion. By making this measurement repetitively with the aid of a lock-in amplifier, the minute signal (0.1 μV or less) can be extracted from the noise; given the resistivity of the medium, the current density can then be readily calculated (for recent developments in vibrating probe technology see Nuccitelli, 1986). Jaffe and his colleagues mapped the currents generated by a variety of cells and organisms including eggs, embryos, amoebae, plant roots and pollen tubes (Jaffe, 1979, 1981; Nuccitelli, 1982). Moreover, they collected persuasive evidence that a spatially localized current of calcium ions is an important element in the process by which germinating pollen grains and algal embryos select the site of outgrowth. This includes the striking demonstration that the calcium current develops prior to outgrowth and predicts its locus (Robinson and Jaffe, 1975; Nuccitelli, 1978; Weisenseel and Kircherer, 1981); and that localized calcium fluxes imposed by the investigator may impress their own polarity upon subsequent outgrowth (Chen and Jaffe, 1979; Robinson and Cone, 1980).

Fungal hyphae grow in a strikingly polarized manner, extending only at the extreme tip. This laboratory began research on the endogenous electric currents of fungi in the hope of learning how currents arise in these organisms, and of gaining fresh insight into the abiding mystery of polarized hyphal growth.

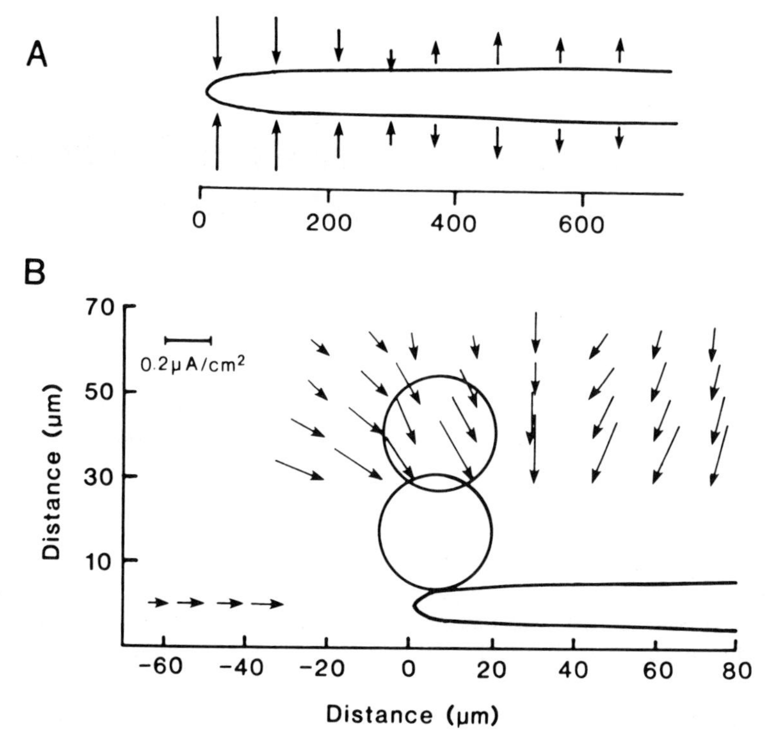

Figure 1. Endogenous electric current generated by a growing hypha of *Achlya bisexualis*: **(A)** Current flows inward in the apical zone, outward beyond 300 μm. **(B)** Detailed map of current flow in the apical region; the circles represent the sensing element of the vibrating probe. The scale bar refers to the length of the vectors. (From Kropf *et al.*, 1984, with permission of the Rockefeller University Press.)

Physical basis of the endogenous current

Figure 1 shows the basic phenomenon; *Achlya bisexualis*, was chosen for this work because of its wide hyphae and moderate rate of hyphal extension. Note that electric current, defined as the flow of positive charge, enters the hyphal tip over a broad zone and exits distally from the hyphal trunk. In aqueous media, charges must be carried by ions. The data imply, therefore, that the hyphal tip is a site of net cation influx or of anion efflux; conversely, there must be net cation efflux from the trunk, or else anion influx. The transport systems that mediate these ion fluxes must be segregated along the hypha. Our first task then, was to identify the ions that carry the electric current and the mechanisms by which they passed across the plasma membrane.

We have devoted considerable effort to these questions and have published many of our findings (Kropf *et al.*, 1984; Gow *et al.*, 1984; Kropf, 1986). Briefly, we believe that the bulk of the electric current is carried by a circulation of protons. Protons are expelled by a proton-translocating ATPase, presumed to belong to the same family as those of *Neurospora* and yeast (Goffeau and Slayman, 1981; Harold, 1986). This generates a proton potential across the plasma membrane, hyphal interior negative and alkaline, that draws protons and other cations back into the cytoplasm. Several ionic fluxes are involved, the chief one being symport of protons with amino acids, which

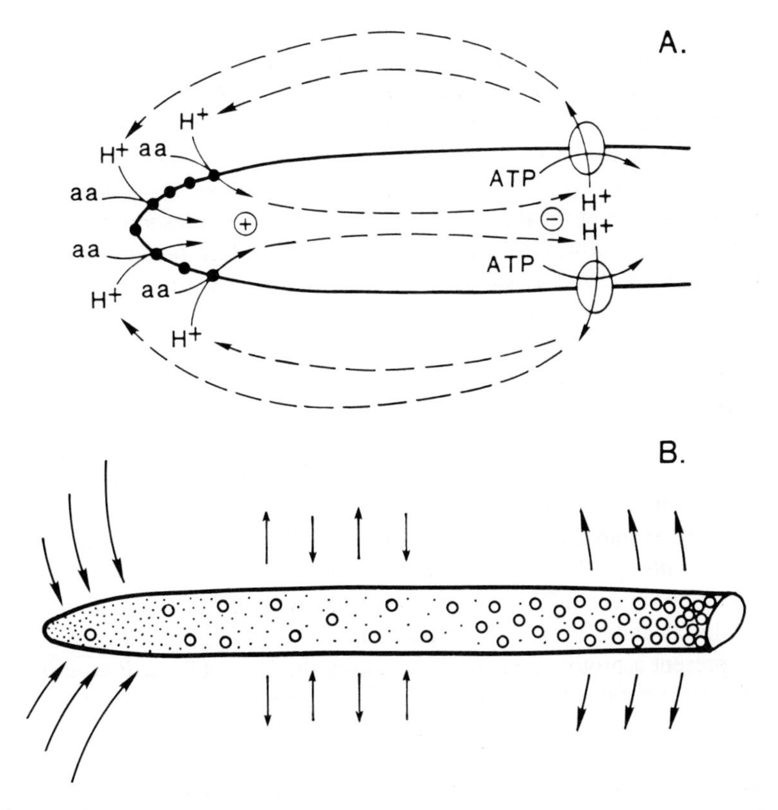

Figure 2. Genesis of the transhyphal proton current, a hypothesis. **(A)** The principle: current arises by the spatial segregation of proton pumps from amino acid/proton symporters. **(B)** Proposed distribution of pumps and porters. Porters (dots) are concentrated towards the hyphal anterior, while pumps (circles) are excluded from the tip and predominate distally. There would be no net current in the zone of overlap.

are *Achlya's* source of sulphur and nitrogen. Methionine transport is especially important, accounting for two thirds of the total current. The transcellular electric current is due to the spatial segregation of the transport systems, as depicted in *Figure 2*: symporters are concentrated in the apical region of the hypha while the ATPase is thought to be situated distally and excluded from the apex. In short, we propose that the transcellular electric current represents a chemiosmotic proton circulation extended in space.

The hypothesis rests on several lines of evidence, all of them necessarily circumstantial. Ion substitution experiments showed that none of the common ions of the growth medium carry significant inward current, leaving protons as the prime candidate. Indeed, the medium adjacent to the apical zone is slightly alkaline while that surrounding the hyphal trunk turns acid, in accord with the proposition that protons flow into the apex and out of the trunk (Kropf *et al.*, 1984; Gow *et al.*, 1984). It is noteworthy that calcium ions are not continuously required for hyphal extension, and make no measurable contribution to the inward current. Removal of amino acids from the medium halts hyphal extension and abolishes the current; both recover when amino acids are restored. Subsequently, Kropf (1986) showed that *Achlya* hyphae generate a large membrane potential, −200 mV or more, apparently by proton extrusion; and that amino acids depolarize the hyphae, as expected for uptake by symport with protons. Recently, we have devised a growth medium that lacks amino acids. The transcellular current under these conditions is minimal, but reappears when the medium is supplemented with amino acids or with certain analogues. In summary, we are quite persuaded that *Achlya* drives a proton circulation by segregating amino acid/proton symporters from the H^+-ATPase. To account for the current, only one class of transport systems need be distributed in a non-uniform manner. We suspect that, in fact, *both* pumps and porters are differentially localized as shown in *Figure 2B*, but have no more than tenuous evidence to support this view.

The degree to which pumps and porters are segregated from one another need not be as large as *Figure 2B* suggests. Slayman and his associates estimate that the H^+-ATPase of *Neurospora* drives protons across the plasma membrane at a rate of 25 $\mu A.cm^{-2}$ (Slayman, 1980). If the pump current of *Achlya* is of similar magnitude, even a small nonuniformity in the placement or the activity of pumps or porters will suffice to generate a transcellular current of 1 $\mu A.cm^{-2}$.

Electric currents into the apex and out of the trunk have been noted in a variety of fungi (Stump *et al.*, 1980; Gow, 1984; Horwitz *et al.*, 1984). The case of *Neurospora* is being examined in detail (Y.Takeuchi, J.H.Caldwell and F.M.Harold, unpublished; A.R.McGillivray and N.A.R.Gow, unpublished). Here again, the electric current appears to represent a proton circulation driven by the H^+-ATPase. Proton influx into the apical zone is not due to symport with amino acids (which this organism does not require), but to the uptake of ions. Symport of inorganic phosphate with protons may be one route of current entry, potassium accumulation another. Calcium ions may also contribute to the inward current, and their uptake is clearly necessary for the apex to extend in the normal manner. Evidently, the transcellular currents of diverse fungi, while alike in spatial pattern, do not necessarily arise in the same manner.

Do ionic currents polarize growth?

Fungal hyphae extend by means of membrane-bound vesicles that supply enzymes and substrates to the apex, where new wall and membrane are produced. The vesicles, thought in Oomycete fungi to be formed in the endoplasmic reticulum and packaged in Golgi bodies, are then transported to the tip in a polarized manner and exocytosed there. The driving force for surface expansion is the turgor pressure. Apical exocytosis permits hyphae to yield locally to this force and generates the characteristic morphology of the hyphal tip (Grove, 1978; Gooday and Trinci, 1980; Koch, 1982; Gooday, 1983; Wessels, 1986). One of the many unanswered questions is, what brings about the extreme polarization of hyphal extension?

The transcellular proton current is itself a manifestation of hyphal polarity, a consequence of the lateral segregation of transport systems as the hypha extends. But one can readily imagine, as L.F.Jaffe has argued in other contexts (Jaffe 1979, 1981), that the ion current helps the hypha to maintain its polarized mode of growth. Currents may exert a morphogenetic effect by more than one mechanism. For example, the influx of protons or calcium ions into the hyphal tip should generate a substantial electric field across the apical cytoplasm, with the apex itself positive. This field in turn may redistribute cytoplasmic organelles or membrane proteins by electrophoresis or electroosmosis and thus help to localize the site of apical exocytosis. The existence of an electric field across the apical cytoplasm of *Neurospora* can be inferred from data published more than two decades ago (Slayman and Slayman, 1962), and recent measurements with *Achlya* indicate a field as steep as 200 $mV.cm^{-1}$ (Kropf, 1986). Alternatively, hyphae may monitor the influx of particular ions, especially Ca^{2+} or protons, and respond by localized extension of the cytoskeleton, by contractile activities or by localized exocytosis. Even an electroosmotic flow of water down the hypha may be part of the process of polarization. There is no shortage of plausible possibilities; the challenge is to devise experiments that are critical but not too simplistic.

An obvious approach would be to inject electric current into a growing hypha and observe the effects on tip extension. We have not yet tried this because the experiments are technically difficult and their interpretation dubious. However, many cells and organisms respond to applied electric fields by altering the direction of movement, growth or development (Lund, 1947; Robinson, 1985). The effect of applied fields on fungal growth, branching and germination (McGillivray and Gow, 1986) are considered in N.A.R.Gow's contribution to this volume. We would emphasize here that applied fields generally exceed the endogenous ones by at least an order of magnitude, so the possibility of artefacts cannot be disregarded. Nevertheless, the common observation that applied fields impose their own polarity upon growth or development without grossly disturbing these delicate processes, has lent credence to the view that ionic or electrical phenomena play a role in the spatial orientation of growth.

Another approach is to alter the growth conditions and look for changes in the pattern of ionic currents. *Figure 3* shows results recently obtained with *Achlya* (W.J.A.Schreurs and F.M.Harold, unpublished). Initially we used medium $DMA_{3.2}$, a rich defined medium containing both glucose and 3.2 mM amino acids. The hyphae extend at a rate of about 5 $\mu m.min^{-1}$, attain a width of about 30 μm, and generate a marked transcellular electric current that flows inward over the apical 300 μm and outward beyond (*Figure 3A*). When the hyphae are grown in a dilute medium ($DMA_{0.06}$), containing only one fiftieth of the usual amino acid complement, they extend at the normal rate but they are thin (10–15 μm) and branch infrequently. The electric current pattern is attenuated and elongated, current entering over the anterior 800–1000 μm with its peak far behind the tip (*Figure 3B*). It seems likely that hyphae respond to the dearth of amino acids by distributing porters over a wider area. Eventually we devised a growth medium that lacks amino acids altogether but contains thioglycolate and urea (1 mM each) as sources of sulphur and nitrogen. The hyphae extend at the normal rate of 5 $\mu m.min^{-1}$, have a width of 10–15 μm, but generate no more than minimal transcellular current (*Figure 3C*). Statistical comparison of the curves at critical points (location and amplitude of peak inward current, and location of the transition from inward current to outward) indicates that the three curves differ significantly at the 5% confidence level. We must conclude that the pattern of transcellular current reflects primarily the rate and spatial localization of amino acid uptake. Since in all three media the hyphae extend at the same rate and in the same polarized manner, it is not likely that apical extension is obligatorily linked to the flow of charge down the hypha. In particular, we cannot believe that the electric current localizes the site of apical exocytosis. This inference reinforces the conclusion reached on other grounds by Kropf *et al.* (1983).

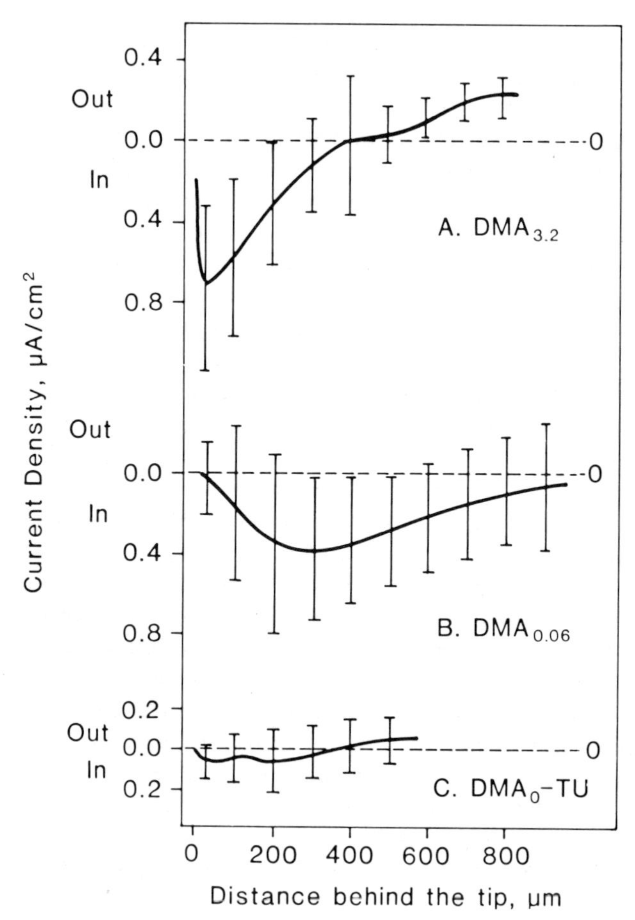

Figure 3. The pattern of electric current depends on the growth medium. **(A)** Hyphae growing in $DMA_{3.2}$. Each point represents the average of measurements on 3−8 individual hyphae, bars the standard deviation. **(B)** Hyphae in $DMA_{0.06}$. Points represent the average of 38−66 hyphae. **(C)** Hyphae in DMA_0-(thioglycolate-urea). Eleven hyphae were mapped.

Surprisingly, when we mapped growing hyphae with extracellular pH microelectrodes we noted that in all three media the region of medium adjacent to the tip was slightly alkaline. We are not quite certain how apical alkalinity arises, but all our observations are consistent with the belief that it reflects the influx of protons into the hyphal tip. The weak electric current seen in thioglycolate-urea medium (*Figure 3C*) may represent a small proton flux that is not coupled to the uptake of amino acids. It appears, then, regardless of the composition of the growth medium, that so long as a hypha extends protons flow into its tip. This, perhaps, is just what one should expect. If it is true that proton pumps are excluded from the apical zone, as *Figure 2B* suggests, then there should always be a net flux of protons into the tip. Proton uptake will be stimulated when amino acids are present, because these are accumulated by symport with protons.

Proton influx will tend to make the apical cytoplasm of hyphae relatively acidic, in accord with cytological observations reported by Turian *et al.* (1985). The magnitude of the proton flux is considerable. Assuming that a proton current of 1 μA.cm^{-2} flows across the surface of a cylinder 100 μm long and 30 μm across, and that the buffer capacity of the cytoplasm equals that of 0.1 M phosphate buffer, we estimate that the cytoplasmic pH would drop by 0.1 unit every 10 min unless the protons are continuously removed. There is no proof that this proton flux serves any physiological function besides energy coupling, but we would find that surprising. What might its role be? The apical region of the hypha into which protons flow is long; we therefore think it unlikely that proton influx determines the site of apical exocytosis. Instead, we suppose that proton influx helps create general conditions suitable for extension: it may supply the protons needed to acidify apical vesicles (Al-Awqati, 1986), modulate the cytosolic calcium level, or perhaps favour reassembly of the cytoskeleton. A tentative hypothesis is that one purpose of the extended proton circulation (*Figure 2*) is to generate this apical zone of proton influx; but whatever the functions of these protons prove to be, we regard them as physiological effectors rather than as spatial signals.

Studies with other fungi provide some support for the proposition that the polarized distribution of pumps and porters inferred above is a general feature of hyphal organization, and that its function may be to generate a flow of protons into the apical zone. Jennings and his associates inferred from cytochemical studies that several fungi exclude ATPase from the tip (Jennings, 1979, and references cited therein). Turian and his associates have repeatedly linked apical extension with acidity (Turian 1981; 1983; Turian *et al.*, 1985). We confess some skepticism regarding the methods employed in these studies, but the gradients they reveal may nevertheless be real. Studies by Y.Takeuchi with *Neurospora* (unpublished) reveal no obligatory coupling between extension rate and the intensity of the electric current, but again link extension to the influx of protons into the tip. It remains for future research to determine whether the correlation between proton influx and hyphal extension is informative and just what it means.

Exocytosis and chemotropic growth

Fungal hyphae extend by the exocytosis of precursor vesicles, a process restricted to the extreme apex. What confines exocytosis to this narrow region? It seems necessary to postulate a highly localized signal of some kind that carries the message 'This is the place'. The nature of this signal is quite unknown. The most popular mechanism calls for a flux of calcium ions, mediated by localized ion channels and closely coupled to each exocytotic event. If these channels were themselves inserted into the plasma membrane by apical vesicles, and closed soon after insertion, exocytosis would be tightly confined to the very tip (Jaffe, 1981; Picton and Steer, 1982; Brawley and Robinson, 1985). This particular mechanism is probably not applicable to *Achlya*, since extension of the hyphae does not require continuous calcium influx and proton influx is not restricted to the extension zone. However, the principle of a short-lived signal carried to the extending tip by the precursor vesicles themselves remains appealing; we have no basis for guessing its molecular identity.

Recent observations do, however, suggest that apical exocytosis may be modulated by nutritional cues. This line of inquiry stems from the observation that, when the growth medium was replaced by one lacking amino acids, hyphal extension ceased immediately and did not usually resume until amino acids were restored (Kropf *et al.*, 1984). *Achlya* hyphae, like those of other fungi, can sometimes extend in the apparent absence of all nutrients, but under our experimental regimen extension is clearly favoured by the presence of organic substances. Most amino acids will serve, some amino acid analogues, even certain metabolites that lack nitrogen including the keto-analogue of methionine and also thioglycolate. Extension, then, is favoured not by nutrients *per se* but by an appropriate signal.

Now, it is well known that in *Achlya* methionine and amino acid mixtures are powerful attractants for chemotropic growth (Musgrave *et al.*, 1977; Manavathu and des S.Thomas, 1985). We have begun to explore the relationship between tip extension and orientation and find good correspondence between these two phenomena. A very tentative hypothesis based on what are still preliminary observations suggests that chemotropic signals are received, not by the H^+/amino acid symporters, but by special receptors confined to the extreme apex (*Figure 4*). It may not be farfetched to imagine that binding of methionine and other amino acids to these receptor proteins orients extension by shifting the locus of apical exocytosis ('bulging growth', in the terminology of Sievers and Schnepf, 1981). Relocation of the site of apical exocytosis, in a manner analogous to that envisaged for nutrient gradients, also provides an alternative explanation for the galvanotropic response to applied electric fields. These notions must now be examined and refined by experiment.

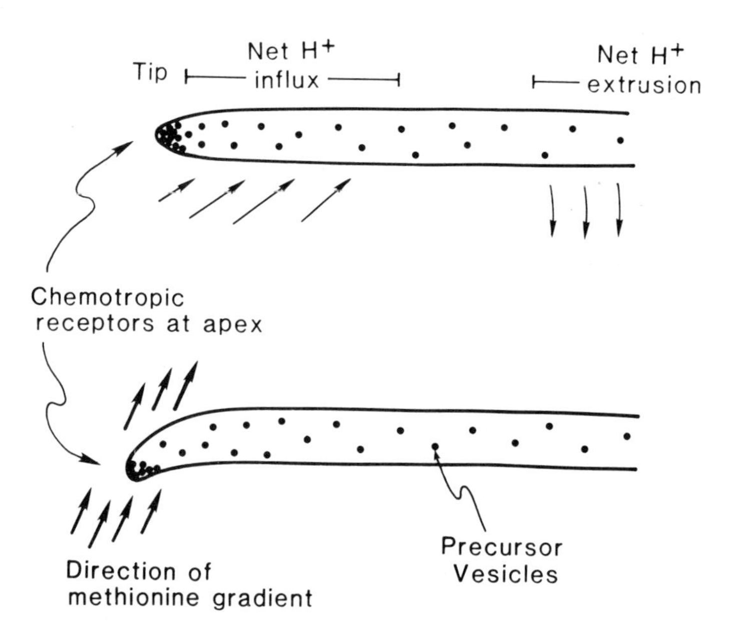

Figure 4. Some hypothetical elements in the polarization of *Achlya* hyphae. **(A)** Influx of protons into the apical region generates an acidic zone that favours extension. Exocytosis of apical vesicles (dots) is autonomous, but modulated by receptor proteins located at the tip. **(B)** The hypha has just encountered a methionine gradient; the gradient displaces the site of exocytosis to favour growth in the direction of the gradient.

Currents and branching

Hyphae extend only at their tips, but when conditions are favourable they put forth branches from previously quiescent regions and thus generate new tips. This must entail the local dissolution of the existing wall. In *Achlya*, whose wall is rich in cellulose, branching is accompanied by the secretion of cellulase from the mycelium; the enzyme appears to be carried in specialized vesicles that accumulate at the site of branch initiation (des S.Thomas and Mullins, 1969; Nolan and Bal, 1974; Mullins and Ellis, 1974). Perhaps turgor pressure forces the weakened wall to bulge and thus gives rise to the branch-initial. This may be followed by migration of nuclei and other organelles into the nascent branch, and by the establishment of apical exocytosis (R.L.Harold, unpublished observations).

We know little of the events that precede emergence of the branch, especially of the factors that determine its location. However, branching was found to be correlated with the establishment of the characteristic ion current associated with an extending tip. Kropf *et al.* (1983) noted that *Achlya* sends out branches from regions of inward current. A new zone of inward current often precedes branching by as much as 30 min, and its locus is near the peak of current entry. There is evidence that the current again represents proton/amino acid symport (Kropf *et al.*, 1984). Very recently, N.A.R.Gow and G.W.Gooday (unpublished) discovered another and most remarkable instance of this phenomenon. Male hyphae of *Achlya* respond to the hormone antheridiol by putting forth specialized sexual hyphae. A new zone of intense inward current precedes and predicts the location of these branches. We would emphasize that in neither instance does the current predict the locus of branch initiation with precision; rather, branches emerge near the peak of a zone of inward current some 200 μm in width, far broader than the branch itself.

It is not at all clear what this phenomenon means. Evidently something takes place in the general area where a branch will later form, and this includes the activation or intensification of ion transport. The connection appears to be a significant one, for ionophores that transport protons or calcium ions induce branching in *Achlya* (Harold and Harold, 1986) and in *Neurospora* (Reissig and Kinney, 1983); cytochalasins do likewise, suggesting that branching involves the assembly of microfilaments. Applied electric fields often stimulate branching and bias the direction of emergence (De Vries and Wessels, 1982; McGillivray and Gow, 1986). We do not believe that genesis of a new zone of ion influx determines the spatial location of the new branch. Rather, we incline to the view that branches emerge from regions of relatively acidic pH; proton influx thus elicits conditions generally favourable to branch initiation, perhaps by a process that involves calcium ions. The actual location of the branch, like that of any other tip, would be determined by additional factors. Seen from this viewpoint, the new zone of inward current bespeaks a loss of hyphal polarity that precedes the formation of a new polar axis, a characteristic phenomenon in developmental biology (we are indebted to B.C.Goodwin for pointing this out to one of us). Evidently, there is something subtle yet to be learned from the frequent observation that ionic currents precede developmental events and predict their location.

Ion currents as consequence and cause of polarity

Ion currents are widespread among eukaryotic organisms, nearly ubiquitous, and should have some general significance. Our experiments have clarified the physical basis of the transcellular currents in *Achlya* and *Neurospora*: each represents a chemiosmotic proton circulation extended in space, arising by the segregation of proton pumps from proton-linked porters (porters apical). Other organisms segregate different sets of transport systems. For example, the germinating pollen grain pumps protons from the grain while K^+ ions leak into the tube (Weisenseel and Jaffe, 1976); in *Pelvetia* embryos, current pulses are due to periodic efflux of chloride from the tip while K^+ ions flow out of the cell body (Nuccitelli and Jaffe, 1976). The first conclusion is simply that cells often place transport systems at particular locations to serve localized functions. One way of fixing transport proteins in place may be to anchor them to the cytoskeleton.

As a rule, we do not know what advantages accrue from the spatial localization of transport systems. In the case of fungal hyphae it may be advantageous to place nutrient porters apically because that is where they will encounter fresh medium as the hypha extends. Most transport systems translocate electric charges for reasons of energetics (Harold, 1986), and when they are spatially segregated within a single compartment electric currents must flow between them. Whether the resulting transcellular ionic current has a physiological function in itself is a separate question, that must be addressed anew for each organism. There is no *a priori* reason to believe that transport systems are always segregated in order to generate a transcellular current. In the case of fungal hyphae, electrical polarity is clearly a manifestation, or consequence, of their polarized organization. Does the flow of protons through the hyphae also play a role in maintaining the polarized mode of extension that generates the ionic current?

We have learned enough to surmise that the answer cannot be a straightforward yea or nay. The striking fact, that hyphae can extend quite normally while driving a minimal electric current through themselves (*Figure 3C*), indicates plainly that the full-blown current pattern (*Figure 1A*) serves no indispensable function. Yet it apparently remains true that so long as a hypha extends, protons flow into its apical zone; and it seems to us intuitively unlikely that so widespread a feature should be only coincidentally associated with the apical mode of hyphal extension. The zone of proton influx is too long to specify the minute site of apical exocytosis. We therefore suspect that proton influx, rather than serving as a mechanism for localizing the tip, plays a role in the physiology of extension. Protons may, for instance, alter the concentration of cytoplasmic calcium ions, which have been implicated in the extension of several kinds of tip-growing organisms (Jaffe *et al.*, 1975; Picton and Steer, 1982, 1983; Hepler and Wayne, 1985; Reiss and Herth, 1985).

How, then, do fungal hyphae maintain their polarized mode of growth? There is probably no single answer, since many of the coordinated activities of growing hyphae are spatially polarized. The structure of the hyphal wall changes progressively as a function of distance behind the tip, microtubules and organelles are axially aligned in ways that may reflect their intrinsic polarity, cytoplasmic flow may have a directional aspect. The flow of protons from the tip toward the trunk takes its place as one of the players in this ensemble, not as the dominant conductor. One especially important aspect of polarization must be whatever confines exocytosis (and thereby the deposition of new wall and membrane) to a tiny region at the extreme apex (Staebell and Soll, 1985). This presumably requires a localized signal, but we remain ignorant of its identity; indeed, we know remarkably little about the mechanism of exocytosis and its regulation in fungi. The manner by which apical exocytosis is localized may be the key to understanding how a hyphal tip continuously reincarnates its form as it forges across the substrate.

Acknowledgements

Ideas expressed herein were generated, challenged and refined in the course of collaboration with Neil Gow, Ruth Harold, Darryl Kropf, Jan Schmid and Yuko Takeuchi. Original research was supported in part by National Institutes of Health grants AI-03568 and GM-33354.

References

Al-Awqati,Q. (1986) Proton-translocating ATPases. *Annual Review of Cell Biology*, **2**, 179–199.

Brawley,S.H. and Robinson,K.R. (1985) Cytochalasin treatment disrupts the endogenous currents associated with cell polarization in fucoid zygotes. *Journal of Cell Biology*, **100**, 1173–1184.

Chen,T.H. and Jaffe,L.F. (1979) Forced calcium entry and polarized growth of *Funaria* spores. *Planta*, **144**, 401–406.

des S.Thomas,D. and Mullins,J.T. (1969) Cellulase induction and wall extension in the water mold *Achlya ambisexualis*. *Physiologia Plantarum,* **22**, 347–353.

DeVries,S.C. and Wessels,J.G.H. (1982) Polarized outgrowth of hyphae by constant electrical fields during reversion of *Schizophyllum commune* protoplasts. *Experimental Mycology*, **6**, 95–98.

Goffeau,A. and Slayman,C.W. (1981) The proton-translocating ATPase of the fungal plasma membrane. *Biochimica et Biophysica Acta*, **639**, 197–223.

Gooday,G.W. (1983) The hyphal tip. In *Fungal Differentiation*. Smith,J.E. (ed.), Marcel Dekker, New York, pp. 315–356.

Gooday,G.W. and Trinci,A.P.J. (1980) Wall structure and biosynthesis in fungi. In *The Eukaryotic Microbial Cell.* Gooday,G.W., Lloyd,D. and Trinci,A.P.J. (eds), *Symposia of the Society of General Microbiology*, **30**, 207–251.

Gow,N.A.R. (1984) Transhyphal electrical currents in fungi. *Journal of General Microbiology*, **130**, 3313–3318.

Gow,N.A.R., Kropf,D.L. and Harold,F.M. (1984) Growing hyphae of *Achlya bisexualis* generate a longitudinal pH gradient in the surrounding medium. *Journal of General Microbiology*, **130**, 2967–2974.

Grove,S.N. (1978) The cytology of hyphal tip growth. In *The Filamentous Fungi*. Smith,J.E. and Berry,D.R. (eds), John Wiley, New York, Vol. III, pp. 28–50.

Harold,F.M. (1986) *The Vital Force: A Study of Bioenergetics*. W.H.Freeman, New York.

Harold,R.L. and Harold,F.M. (1986) Ionophores and cytochalasins modulate branching in *Achlya bisexualis*. *Journal of General Microbiology*, **132**, 213–219.

Hepler,P.K. and Wayne,R.O. (1985) Calcium and plant development. *Annual Review of Plant Physiology*, **36**, 397–439.

Horwitz,B.A., Weisenseel,M.H., Dorn,M.H. and Gressel,J. (1984) Electric currents around growing *Trichoderma* hyphae, before and after photoinduction of conidiation. *Plant Physiology*, **74**, 912–916.

Jaffe,L.F. (1979) Control of development by ionic currents. In *Membrane Transduction Mechanisms*. Cone,R.A. and Dowling,J.E. (eds), Raven Press, New York, pp. 199–231.

Jaffe,L.F. (1981) The role of ionic currents in establishing developmental patterns. *Philosophical Transactions of the Royal Society of London B*, **295**, 553–566.

Jaffe,L.F. and Nuccitelli,R. (1974) An ultrasensitive vibrating probe for measuring steady electrical currents. *Journal of Cell Biology*, **63**, 614–628.

Jaffe,L.A., Weisenseel,M.H. and Jaffe,L.F. (1975) Calcium accumulations within the growing tip of pollen tubes. *Journal of Cell Biology*, **67**, 488–492.

Jennings,D.H. (1979) Membrane transport and hyphal growth. In *Fungal Walls and Hyphal Growth.*

Burnett,J.H. and Trinci,A.P.J. (eds), Cambridge University Press, Cambridge, pp. 279–294.

Koch,A.L. (1982) The shape of the hyphal tip of fungi. *Journal of General Microbiology*, **128**, 947–951.

Kropf,D.L. (1986) Electrophysiological properties of *Achlya* hyphae: Ionic currents studied by intracellular potential recording. *Journal of Cell Biology*, **102**, 1209–1216.

Kropf,D.L., Lupa,M.D.A., Caldwell,J.H. and Harold,F.M. (1983) Cell polarity: ion currents precede and predict branching in the water mold *Achlya. Science*, **220**, 1385–1387.

Kropf,D.L., Caldwell,J.H., Gow,N.A.R. and Harold,F.M. (1984) Transcellular ion currents in the water mold *Achlya*. Amino acids proton symport as a mechanism of current entry. *Journal of Cell Biology*, **99**, 486–496.

Lund,E. (1947) *Bioelectric Fields and Growth.* University of Texas Press, Austin, Texas, USA.

Manavathu,E.K. and des S.Thomas,D. (1985) Chemotropism of *Achlya bisexualis* to methionine and methionyl compounds. *Journal of General Microbiology,* **131**, 751–756.

McGillivray,A.M. and Gow,N.A.R. (1986) Applied electrical fields polarize the growth of mycelial fungi. *Journal of General Microbiology,* **131**, 751–756.

Mullins,J.T. (1973) Lateral branch formation and cellulase production in the water moulds. *Mycologia*, **65**, 1007–1014.

Mullins,J.T. and Ellis,E.A. (1974) Sexual morphogenesis in *Achlya*: Ultrastructural basis for the hormonal induction of antheridia. *Proceedings of the National Academy of Sciences USA*, **71**, 1347–1350.

Musgrave,A., Ero,L., Scheffer,R. and Oehlers,E. (1977) Chemotropism of *Achlya bisexualis* germ hyphae to casein hydrolysate and amino acids. *Journal of General Microbiology*, **101**, 65–70.

Nolan,R. and Bal,A.K. (1974) Cellulase localization in hyphae of *Achlya bisexualis. Journal of Bacteriology*, **117**, 840–843.

Nuccitelli,R. (1978) Öoplasmic segregation and secretion in the *Pelvetia* egg is accompanied by a membrane generated electrical current. *Developmental Biology*, **62**, 13–33.

Nuccitelli,R. (1982) Transcellular ion currents: signals and effectors of cell polarity. *Modern Cell Biology*, **2**, 451–481.

Nuccitelli,R. (ed) (1986) *Ionic Currents in Development*. A.R.Liss, New York.

Nuccitelli,R. and Jaffe,L.F. (1976) The ionic components of the current pulses generated by developing fucoid eggs. *Developmental Biology*, **49**, 518–531.

Picton,J.M. and Steer,M.W. (1982) A model for the mechanism of tip extension in pollen tubes. *Journal of Theoretical Biology*, **98**, 15–20.

Picton,J.M. and Steer,M.W. (1983) Evidence for the role of Ca^{2+} ions in tip extension in pollen tubes. *Protoplasma*, **115**, 11–17.

Reiss,H.D. and Herth,W. (1985) Nifedipine-sensitive calcium channels are involved in polar growth of lily pollen tubes. *Journal of Cell Science*, **76**, 247–254.

Reissig,J.L. and Kinney,S.G. (1983) Calcium as a branching signal in *Neurospora crassa. Journal of Bacteriology*, **154**, 1397–1402.

Robinson,K.R. (1985) The responses of cells to electrical fields: A review. *Journal of Cell Biology*, **101**, 2023–2027.

Robinson,K.R. and Cone,R. (1980) Polarization of fucoid eggs by a calcium ionophore gradient. *Science*, **202**, 77–78.

Robinson,K.R. and Jaffe,L.F. (1975) Polarizing fucoid eggs drive a calcium current through themselves. *Science*, **187**, 70–72.

Sievers,A. and Schnepf,E. (1981) Morphogenesis and polarity of tubular cells with tip growth. In *Cytomorphogenesis in Plants. Cell Biology Monographs.* Kiermayer,O. (ed.), Springer Verlag, Vienna, Vol. 8, pp. 265–299.

Slayman,C.L. (1980) Transport control phenomena in *Neurospora*. In *Plant Membrane Transport: Current Conceptual Issues.* Spanswick,R.M., Lucas,W.J. and Dainty,J. (eds), Elsevier/North Holland, Amsterdam, pp. 179–180.

Slayman,C.L. and Slayman,C.W. (1962) Measurement of membrane potentials in *Neurospora. Science*, **136**, 876–877.

Staebell,M. and Soll,D.R. (1985) Temporal and spatial differences in cell wall expansion during bud and mycelium formation of *Candida albicans. Journal of General Microbiology*, **131**, 1467–1480.

Stump,R.F., Robinson,K.R., Harold,R.L. and Harold,F.M. (1980) Endogenous electrical currents in the water mold *Blastocladiella emersonii* during growth sporulation. *Proceedings of the National Academy of Sciences USA*, **77**, 6673–6677.

Turian,G. (1981) Decreasing pH-gradient toward the apex of germinating pollen tubes. *Botanica Helvetica*, **91**, 161–167.

Turian,G. (1983) Polarized acidification at germ tube outgrowth from fungal spores (*Morchella* ascospores, *Neurospora* conidia). *Botanica Helvetica*, **93**, 27–32.
Turian,G., Ton-That,T.C. and Perez,R.O. (1985) Acid tip linear growth in fungi: requirements for H^+/Ca^{2+} inverse gradients and cytoskeleton integrity. *Botanica Helvetica*, **95**, 311–322.
Weisenseel,M.H. and Jaffe,L.F. (1976) The major growth current through lily pollen tubes enters as K^+ and leaves as H^+. *Planta*, **133**, 1–7.
Weisenseel,M.H. and Kircherer,R.M. (1981) Ionic currents as control mechanisms in cytomorphogenesis. In *Cytomorphogenesis in Plants. Cell Biology Monographs*. Kiermayer,O. (ed.), Springer Verlag, Vienna, Vol. 8, pp. 379–399.
Wessels,J.G.H. (1986) Cell wall synthesis in apical hyphal growth. *International Review of Cytology*, **104**, 37–79.

CHAPTER 3

Polarity and branching in fungi induced by electrical fields

NEIL A.R.GOW

Department of Genetics and Microbiology, University of Aberdeen, Marischal College, Aberdeen AB9 1AS, UK

Introduction

There has been a recent renaissance of interest in bioelectricity in cell biology. This is truly a rebirth of ideas and observations and not a new perspective of things since experiments on biological electricity and the application of electrical fields to cells date back over two centuries (Bertholon, 1783). In Elmer Lund's remarkable book 'Bioelectric Fields and Growth' which was published in 1947 he outlines concepts, experimental methodologies, data and unanswered questions which are still the focus of attention for those of us who have chosen to look again at this subject. Lund lists over 1400 references; testament to a time when electricity was a preoccupation of the scientific community, the most powerful and useful of forces and a force it seemed that biological systems surely must have learnt to make use of.

Within Lund's lifetime, however, interest in bioelectricity dwindled at the expense of more profitable work in which the chemical and molecular dissection of biology came to the fore. Two problems were responsible for the dulling of interest and consequent loss of momentum in this area. The problems lay not in whether living cells generate electrical currents and electrical fields, because they certainly do, nor whether they are affected by electrical fields, because they clearly are. The problems are firstly in answering whether the electrical phenomena are of any importance. Are they important for cell development, for example, or for cell polarity, or directional and orientated cell growth as has been claimed. Secondly, if they do affect any or all of these fundamental processes how does the electrical activity or responsiveness of a cell articulate with other physiological processes which are collectively responsible for these activities?

We have, however, moved forward recently in a number of ways. We have better and more accurate equipment for measuring the electrical currents of cells. We can profit from the insights yielded by the studies of electrophysiology, bioenergetics and genetics and we have improved techniques for studying the structural organization of cells. In the more recent work which is the subject of this review there has been much gained by the availability of these new techniques. Of these the development of the vibrating probe, an instrument for the detection of minute electrical fields in the extracellular space around the surface of cells, has been invaluable. To this, and many of the new concepts in this area, we owe much to Lionel Jaffe who along with Richard Nuccitelli developed and now market the vibrating probe. I will not describe this instrument since a description is outlined in the previous chapter and the basic design and subsequent modifications and variations have been published. (Jaffe and Nuccitelli, 1974; Dorn and Weisenseel, 1982; Nawata, 1984; Freeman *et al.*, 1985; Nuccitelli, 1986). The vibrating probe has made studies of the electrical activities of fungi possible since the extracellular electrical fields ($1-50$ nV.μm^{-1}) and electrical currents ($0.05-2.5$ $\mu A.\mu m^{-2}$) generated by fungal hyphae are too small to be measured with conventional electrodes.

The growth associated electrical currents of fungi and in particular that of the water-mould *Achlya bisexualis* are amongst the best studied of all the bioelectrical currents that have been investigated. The work which has led to the characterization of the endogenous current in this organism is described in the preceding chapter. The relative simplicity of cell structure and ease of cultivation make fungi excellent organisms for studying bioelectricity and its significance to cell growth. In the fungal hyphae we are seeking to determine whether the transcellular electrical current (ion current or ionic current) is an effector of positional information. That is, we want to know whether the electrical current is something to do with the way in which hyphae localize growth only at the extreme hyphal apex. We know that cell expansion is restricted to the tapering region or so called extension zone at the tip and that growth of the hyphal tip is supported by membrane vesicles which are synthesized in the Golgi apparatus or endoplasmic reticulum in the distal hypha and are then transported to the apex where they fuse by a process akin to exocytosis. The vesicles contain biosynthetic enzymes such as chitin synthetase (Bartnicki-Garcia *et al.*, 1979; Gooday, 1983) and so their fusion provides the enzymes for tip growth and the new plasmalemma for apical expansion. For tip growth to be possible these vesicles must therefore be transported to the apex faster than the rate at which the tip is extending and secondly they must be allowed to fuse only at this tapering apical zone. Much of the impetus for the work of fungal bioelectricity has been directed to answering whether the electrical currents determine or influence either or both of these processes. Before discussing how studies using applied electrical fields have contributed to the investigation of these phenomena, it is of use to summarize the information that has implicated electrical currents and fields in the establishment and maintenance of cell polarity in fungal hyphae and other types of polarized cells.

Correlation between electrical and morphological polarity

Essentially five types of experimental observations suggest a correlation between the electrical activities of tip growing cells such as fungal hyphae and their polarity.

(i) When a profile of electrical current is mapped around a hypha using a vibrating probe it is found that the current is symmetrically orientated around the axis of growth (*Figure 1*). Since a current is defined as the movement of positive charges this means that there is a net uptake of cations or net efflux of anions around the growing tip and a net efflux of cations or uptake of anions in the non-extending region further back. Consequently the tip would be expected to be partially

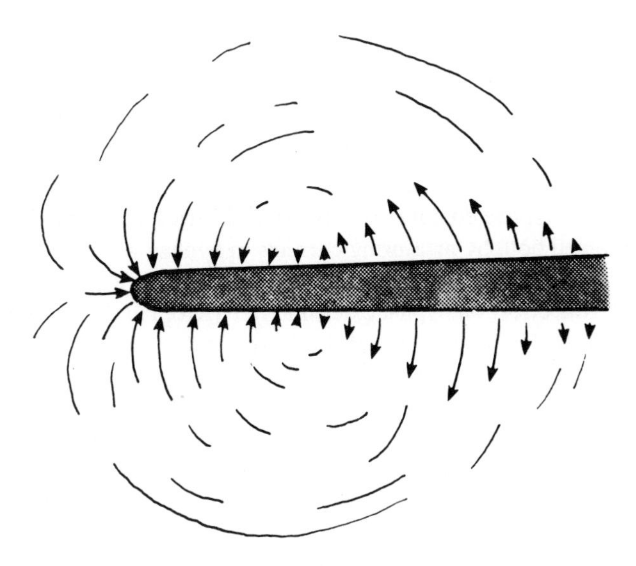

Figure 1. Pattern of transhyphal ion/electric current around fungal hyphae. From Gow (1984), with permission.

depolarized due to the influx of positive charge. This symmetrical profile is shared by all the filamentous fungi which have so far been examined including *Achlya debaryana* (Armbruster and Weisenseel, 1983), *Achlya bisexualis* (Kropf *et al.*, 1983), *Achlya ambisexualis* (Gow and Gooday, in preparation), *Mucor mucedo, Neurospora crassa, Aspergillus nidulans, Schizophyllum commune* and *Coprinus cinereus* (Gow, 1984), *Trichoderma harzianum* (Horwitz *et al.*, 1984), *Basidiobolus ranarum* and *Phytophthora frageriae* (Shand, McCulloch and Gow, unpublished).

(ii) Endogenous currents and associated electrical fields flow in the same way for all tip growing organisms. Positive current always enters the growing tip. This is the case for all of the fungi mentioned above and for plant roots and their root hairs (Weisenseel *et al.*, 1979; Miller *et al.*, 1986), germinating eggs of brown and red algae (Jaffe, 1966; Robinson and Jaffe, 1975), lily pollen germ tubes (Weisenseel *et al.*, 1975) the rhizoids of *Blastocladiella emersonii* (Stump *et al.*, 1980) and rhizoids and repair shoots of red alga *Griffithsia pacifica* (Waaland and Lucas, 1984), the growing ends of carrot embryos (Brawley *et al.*, 1984) and the axons and dendrites of nerve cells (Freeman *et al.*, 1985). The carrier ions for the inward electrical current varies; protons for *A. bisexualis* (Kropf *et al.*, 1984; Gow *et al.*, 1984), *N. crassa* (McGillivray and Gow, in press) and barley roots (Weisenseel *et al.*, 1979), potassium for lily pollen germ tubes (Weisenseel and Jaffe, 1976) and calcium for *Pelvetia* eggs (Robinson and Jaffe, 1975). It is the vectorial flow of electrical charge which is conserved in these widely varying systems. Inward current therefore correlates with polarized growth. It should be noted, however, that conditions have been found in which inward current and cell extension are dissociated. For example, it has been reported that the apical current of a branching hypha of *A. bisexualis* sometimes reverses transiently yet extension continues unabated (Kropf *et al.*,

1983). Despite the reversed charge flow, protons probably continue to enter the apex (Gow *et al.*, 1984). Rarely, outward currents have been observed at the tips of growing hyphae of *N. crassa* (McGillivray and Gow, in preparation). Finally, Waaland and Lucas (1984) report localized growth of rhizoids of *G. pacifica* in the absence of any current.

(iii) Non-growing hyphae do not normally drive electrical currents. This correlation has been noted both by measuring currents in hyphae which had stopped growing naturally (Gow, 1984) or by turning off the electrical current specifically and noting the effect on hyphal extension rate (Kropf *et al.*, 1984). In the latter the ion current flowing into tips of *A. bisexualis* was shown to be due to localized amino acid/proton symport in this region. Depletion of amino acids in the medium or reduction in the extracellular proton concentration (achieved by increasing pH) both turned off current and hyphal extension simultaneously. Restoration of amino acids and the original pH restored current flow and tip growth. Ion currents have, however, been observed around non-growing rhizoids of *G. pacifica* (Waaland and Lucas, 1984) and non-growing hyphae of *N. crassa* (McGillivray and Gow, in press).

(iv) If electrical currents are in some way involved in establishing and then maintaining cell polarity they must exist at or before the time of morphological signs of new cell growth or changes in the pattern of growth. It has been found in a number of systems that new currents and changes in currents precede and predict the production of new growing points or the pattern of growth in an established growing system. For example Kropf *et al.* (1983) showed that the production of vegetative branches of *A. bisexualis* was preceded by an inwardly directed ionic current into the hyphal trunk at a site that predicted where the branches emerged. Recently Gow and Gooday (in press) showed that when male hyphae of *A. ambisexualis* were treated with the female sex pheromone antheridiol, a number of events ensued which correlated with the electrical activity of the stimulated hyphae. Firstly, the inward current at the apex declined as hyphae slowed down and eventually stopped. Sometime afterwards peaks of inward current developed along the hyphal trunk. These increased in size and eventually sexual antheridial branches emerged at the positions of maximum inward current (*Figure 2*). The future sites of outgrowth of the rhizoids of *Pelvetia* zygotes (Nuccitelli, 1978) and pollen tubes of lily plants (Weisenseel *et al.*, 1975) are also predicted by a localized inflow of electrical current.

Gravitropic bending of roots has been investigated by Behrens *et al.* (1982). Vertically grown roots exhibited a symmetrical current pattern which entered the root cap, apical meristem and the beginning of the elongation zone. When vertically grown roots were tilted to the horizontal position the current pattern shifted within 30 seconds so that more current flowed into the upper surface of the root. Stimulated cell division and elongation resulted in visible gravitropism some ten minutes later. The changes in the direction and magnitude of the root currents preceded bending and correlated with the bending pattern suggesting there was a link between the current and gravitropism. No equivalent correlation between bending and current profile has been recorded for fungal hyphae.

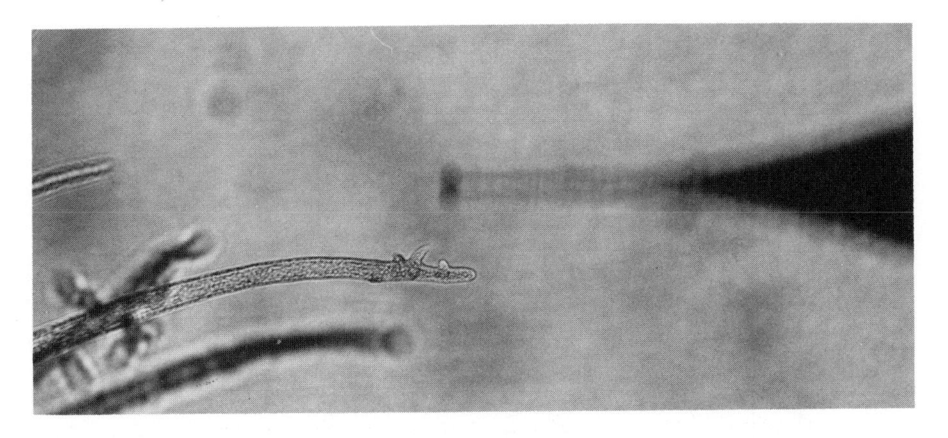

Figure 2. Vibrating probe positioned above sexual branches of *A. ambisexualis* that were induced with antheridiol.

However, electrical events predict changes in the direction of cell growth and the production of new growing points.

(v) The final class of evidence linking the endogenous electrical activity of cells to their polarity has come from experiments in which exogenous electrical fields have been applied to cells and the consequence to the polarity of the cells monitored. The rationale for these experiments is to test whether artificial fields impose a sense of polarity on growing cells. This approach would seem to be of particular significance since most of the other types of evidence are essentially correlative. The fact that electrical currents accompany the establishment and maintenance of cell polarity need not imply that they are causal in this respect. If, however applied fields affect cells this implies that endogenous electrical fields may also be capable of affecting cell polarity. Stretching the argument to its limit the artificial electrical fields may mimic the effects of natural ones either by forcing a current through a cell or causing the redistribution of proteins in the cell or cell membrane. There are, however, inherent difficulties in interpreting these experiments and these will be discussed later.

Analogous experiments have been carried out where induced assymetries in the ionic composition of cells is brought about using gradients of ions or ionophores, particularly for those ionic species which actually carry natural currents. Endogenous and exogenous electrical fields may establish stable ion gradients in the cytoplasm and so these experiments suggest alternative ways in which voltage gradients may generate polarity.

Effects of applied d.c. electrical fields

Experiments applying artificial electrical fields are relatively simple to perform but care needs to be taken to avoid possible artefacts caused by changes in local temperature or pH, the build up of electrode products or the production of gradients of pH, nutrients or oxygen tension. It is also important to establish the lowest fields capable of influencing the cells since we seek to understand how natural electrical fields generate polarity

Table 1. Studies of effects of electrical fields on living cells.

Cell type	*Response*	*References*
Animal cells		
Xenopus neurite	increased growth towards cathode	Hinkle *et al.*, 1981; McCaig, 1986; Patel and Poo, 1982, 1984
Xenopus neural crest	migrate towards cathode; perpendicular alignment	Stumpf and Robinson, 1983; Cooper and Keller, 1984
Axolotyl neural crest	migrate towards cathode; perpendicular alignment	Cooper and Keller, 1984
Quail neural crest	migrate towards cathode; perpendicular alignment	Erickson and Nuccitelli, 1982
Quail fibroblasts	migrate towards cathode; perpendicular alignment	Erickson and Nuccitelli, 1984; Nuccitelli and Erickson, 1983
Fish scale keratocytes	migrate towards cathode	Cooper and Schliwa, 1985
Xenopus myoblasts	migrate towards cathode; perpendicular alignment	Hinkle *et al.*, 1981
Xenopus embryo epithelia	migrate towards cathode; perpendicular alignment	Luther *et al.*, 1983
Mouse embryo fibroblasts	perpendicular alignment	Onuma and Hui, 1985
Plant cells		
Fucus inflatus zygote	rhizoids formed towards anode	Lund *et al.*, 1923
Ulva zygotes	rhizoids formed towards anode	Sand, 1973
Funaria species	rhizoids formed towards anode	Chen and Jaffe, 1979
Equisetum species	growth towards anode	Bentrup, 1968
Zea roots	bending towards anode	Miller and Gow, unpublished
Avena coleoptile	bending towards anode	Schrank, 1947

Table 1. continued.

Cell type	*Response*	*References*
Hydractinia stolon	branching induced by positive current	Müller and El-Shershaby, 1981
Fucus serratus	rhizoids formed towards cathode	Bentrup, 1968
Vinca pollen tubes	emergence and growth towards cathode	Marsh and Beams, 1945
Micrasterias	growth towards cathode	Brower and McIntosh, 1980
Obelia	regeneration towards cathode	Lund, 1947
Nicotinia callus	stimulated growth. Shoot formation towards anode and cathode	Rathore and Goldsworthy, 1985
Pelvetia zygotes	rhizoids formed towards anode, cathode or both with different batches of eggs and varying field strengths	Peng and Jaffe, 1976
Fungal cells		
Schizophyllum commune	protoplasts regenerate towards anode	De Vries and Wessels, 1982
Achlya bisexualis	growth and branching towards anode	McGillivray and Gow, 1986
Neurospora crassa	germination, growth and branching towards anode perpendicular growth at high field strengths	McGillivray and Gow, 1986
Sordaria macrospora	growth and branching towards anode	Gow, Robertson and Read, unpublished
Aspergillus nidulans	growth and branching towards cathode	McGillivray and Gow, 1986
Mucor mucedo	growth and branching towards cathode	McGillivray and Gow, 1986
Rhizopus stolonifer	growth towards cathode	Clark and Goldsworthy, unpublished
Trichoderma harzianum	growth towards anode or cathode or perpendicular according to field strength	McGillivray and Gow, 1986
Phytophthora parasitica	zoospores swim towards cathode	Troutman and Wills, 1964
7 other *Phytophthora* zoosporic species	faster swimming towards the anode	Khew and Zentmeyer, 1971

and these natural fields are generally weak. For most types of cells the physiological range of steady electric fields would be in the order of 0.1 – 10 mV per cell diameter. For fungi with hyphae and spores of about 10 μm diameter this means that fields of less than 10 V.cm^{-1} are of most interest. The smaller the field that is capable of modulating growth the greater the confidence that the experiment is meaningful. Careful control of the experimental conditions and quantification of the effects of applied fields are therefore essential. Over the years there have been many studies of the effects of artificial electrical fields and in most cases the results have been striking (*Table 1*). Animal, plant and fungal cells have been shown to migrate, swim, bend and grow towards one electrical pole or the other. Animal cells capable of migrating do so predominantly towards the cathode (Robinson, 1985). At higher field strengths (above 4 V.cm^{-1}) many cell types become progressively elongated in shape and they migrate with their long axis perpendicular to the electric field (*Table 1*). Growing nerve cells exposed to electric fields bent towards the cathode, produced more filopodia at the growth cone and extended at a rate that was up to eight times faster when they were cathode facing as compared to anode facing (Jaffe and Poo, 1979; McCaig, 1986).

Plant cells and tissues show no common response to electrical fields. Some germinate and grow towards the anode and others do so towards the cathode (*Table 1*). The fertilized egg cells of the alga *Fucus inflatus* (Lund, 1923) are polarized to the anode, while those of *Fucus serratus* grow to the cathode (Bentrup, 1968). From sixteen batches of eggs of the related alga *Pelvetia vestigiata*, eleven initiated rhizoid formation towards the positive electrode, two towards the negative electrode and three towards the negative electrode in weak fields and towards the positive pole in stronger ones (Peng and Jaffe, 1976). These results raise a problem which also occurs when considering the responses of mycelial fungi to electrical fields. The endogenous electrical field of Fucoid eggs such as *P. fastigiata* and *Fucus vesiculosus* is orientated in the same direction and it is presumed that positive current always flows into the rhizoidal end of all algal zygotes (Robinson and Jaffe, 1975; Nuccitelli, 1978: Nuccitelli and Jaffe, 1974). Thus in Fucoid eggs there is no correlation between the profile of the endogenous field and the response to an applied one. A range of other plants including the roots and coleoptiles of higher plants have been shown to show galvanotropism and stimulated growth in an electrical field (*Table 1*; Lund, 1947). All of these plant cell and tissue types showed significant polarization at field strengths of 10 mV per cell diameter although in many cases complete polarization of a population of cells could only be achieved by very much larger fields. Rathore and Goldsworthy (1985) showed that weak negative current in the order of 1 microamp stimulated the growth of tobacco callus while positive current retarded it. The application of the current also induced shoot regeneration from the callus (Rathore, 1986). These studies of plant systems in electric fields show that complex tissues as well as individual cells can exhibit galvanotropism. Again growth rate as well as the direction of growth can be affected by electric fields.

The effects of applied electrical fields on fungi have been examined recently. De Vries and Wessels (1982) devised a simple apparatus which made use of a flat-bed DNA-electrophoresis apparatus to apply an electrical field to a population of protoplasts of *Schizophyllum commune* which were embedded in an agarose gel. Growth medium

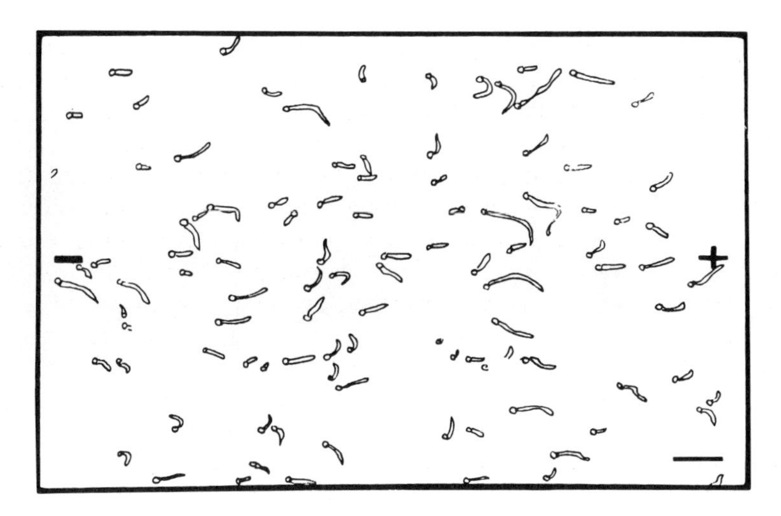

Figure 3. Tracing of an agarose gel in which conidia of *N. crassa* have been allowed to germinate in a field of 30 V cm^{-1} showing polarization towards the anode.

was pumped over the gel and rapidly recirculated between the electrode reservoirs in order to prevent gradients of nutrients or electrode products from building up. They showed that the point of outgrowths of regenerating protoplasts were orientated towards the anode so that a polarization level of 10% was achieved by fields as low as 0.7 mV per cell. The maximum polarization achievable was 75% in a field of 25 mV per cell. De Vries and Wessels also showed that the incorporation of radiolabelled *N*-acetyl glucosamine into chitin in the regenerating wall was not polarized before the hyphae emerged suggesting that the electric field did not affect the distribution of chitin synthetase. If applied fields affect the mobilization and localization of perimembranous proteins (discussed later) this suggests proteins other than wall synthesizing enzymes are responsible for the primary events which define what will become a tip.

Using a similar experimental system McGillivray and Gow (1986) described the effects of applied fields on five mycelial fungi. Germination of conidia of *N. crassa* was polarized towards the anode and occurred more quickly and more synchronously in applied fields (*Figures 3* and *5a*) than normally. Branch formation was stimulated 4-fold by a field of 40 mV per cell and branches were again formed preferentially towards the anode (*Figures 4a* and *5b*). Once a hyphal tip had been formed it also bent towards the anode in fields of less than 10 mV per cell. At higher field strengths there was an increasing tendency for established hyphae to grow perpendicularly to the field (*Figure 4b*) with new branches continuing to be formed on the anode facing side of the perpendicular hyphae. This situation is therefore analogous to the galvanotactic behaviour of animal cells in strong fields in which the cells align perpendicularly and produce growth processes from the leading edge of the cell (*Table 1*). Perpendicular alignment may result if continued anodotropic growth becomes inhibitory, perhaps due to large perturbations of the membrane potential induced by the electric field (Cooper and Keller, 1984; McGillivray and Gow, 1986). The threshold fields strengths for the polarization

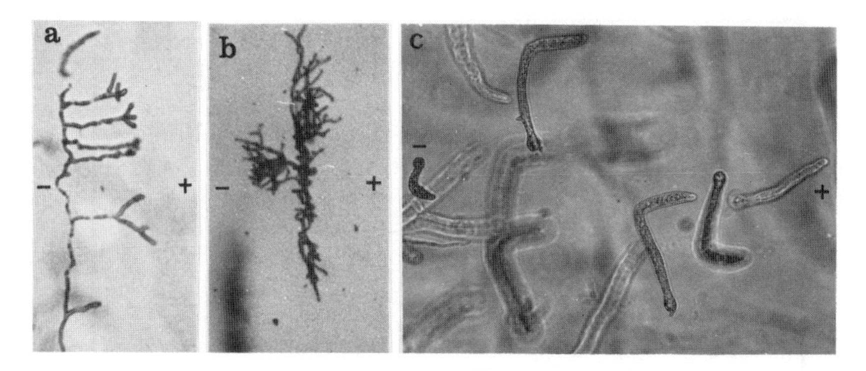

Figure 4. Growth of colonies of *N. crassa* (**a,b**) and *A. bisexualis* (**c**) in electric fields of 30 V cm^{-1} (**a,b**) and 15 V cm^{-1} (**c**) showing branching towards the anode (**a**) perpendicular growth (**b**) and anodotropism (**c**). (**b**) from McGillivray and Gow (1986) with permission.

of germination, hyphal extension and branching were around 1 mV per cell diameter or 1 V.cm^{-1}. More than 90% of hyphae grew anodotropically in fields of 5 mV per cell. Hyphae and branches of *A. bisexualis* were also anodotropic (*Figure 4c*) and hyphae were seen to branch occasionally from their apices, an event which never occurred in controls. The hyphae were otherwise normal morphologically. As noted previously for experiments on plants, different fungi, even closely related species, grew towards different electrodes. For example, *Aspergillus nidulans* like *N. crassa* is an ascomycete yet *A. nidulans* exhibited tropisms towards the cathode. *Mucor mucedo* and *Rhizopus stolonifer* also showed preferential outgrowth towards the negative electrode (*Table 1*). *Trichoderma harzianum* hyphae tended to grow cathodotropically but branches were produced predominantly towards the anode. Motile zoospores of various plant pathogenic *Phytophthora* species have been shown to swim to one or other electrode of applied fields. This might be of importance to their ability to find and infect their host root tips since roots generate electrical fields which are large enough to cause galvanotaxis (Troutman and Wills, 1964; Miller, Shand and Gow, in preparation). Fungi therefore are like plants in exhibiting tropisms to both the positive and negative electrodes. Again it is noted that the endogenous electrical currents and fields of all fungal hyphae that have been examined have the same polarity and so there is no simple correlation between the endogenous electric field and the response to an applied one. Thus, voltage gradients influence polarity rather specifically; growth *per se* was not inhibited therefore all other anabolic, catabolic and cell cycle processes were presumably normal. The more difficult issue is in evaluating the contribution that these data make to our understanding of the control of cell polarity. We need to establish how an electrical field might bring about polarized growth.

Physiological consequences of electric fields

Measurement and mapping of natural electrical fields and currents that are associated with polarized growth and observation of the polarizing influence of applied electric fields suggest that there is an electrical dimension to the control of cell polarity. However, we do not fully understand how these phenomena affect the spatial organization of

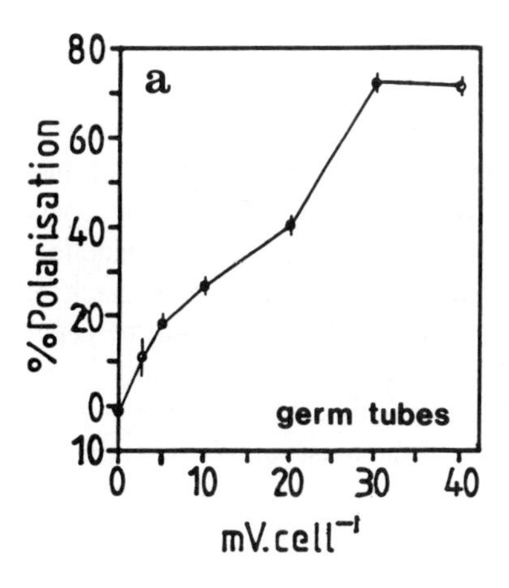

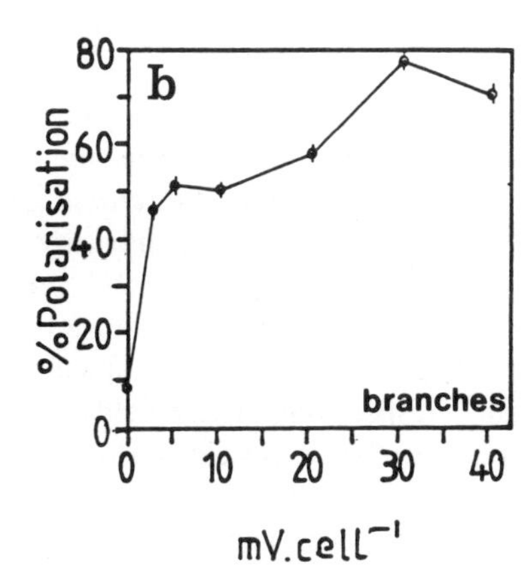

Figure 5. Polarization of germ tubes (**a**) and branches (**b**) in applied dc electrical fields. The percentage polarization was determined as the average cosine that germ tubes or branches made to the anode cathode axis, times one hundred. Zero percent represents random orientation and all positive values orientation towards the anode. From McGillivray and Gow (1986) with permission.

eukaryotic cells. In trying to evaluate the physiological significance and consequences of electrical fields, two approaches can be used. Models of electrically active cells such as fungal hyphae can be constructed, into which data describing the sizes of the electrical fields ionic fluxes, diffusibility and electrophoretic mobility of ions, proteins and organelles can be incorporated. These theoretical considerations provide us with ideas of what physiological effects are at least possible. Complementary experimental studies can show whether electric fields move proteins within the cell or cell membrane or generate standing ion gradients in the cytoplasm. Both approaches fall short of identifying proteins, enzymes, ions or forces which bring about polarity *in vivo* but they provide insights which will help guide further experimentation. It must be borne in mind that polarity may be a property of an integrated system consisting of electrical, chemical, osmotic and structural elements. It may be neither sensible nor meaningful to try to implicate one factor as the sole cause of polarity nor any one system for all organisms exhibiting polarity (see Discussion of Goodwin in this volume). At least two physiological consequences of exogenous and endogenous electric fields can be envisaged: (i) they might mobilize and localize charged proteins or organelles in the cytoplasm or cell membrane either by electrophoresis or electroosmosis (Jaffe *et al.*, 1974; Jaffe, 1977, 1981) or, (ii) they may drive an ion current through the cell and establish ionic gradients in the cytoplasm.

Electrical consequences

A natural ion current flowing through a cell will generate an internal voltage gradient the magnitude of which will be determined by the current density and the cytoplasmic resistivity. Applied electric fields may generate similar intracellular voltage gradients by depolarizing the membrane at the cathodic end of a cell and hyperpolarizing it at the anodic end. Slayman and Slayman (1962) inserted microelectrodes into the apices and distal regions of hyphae of *N. crassa* and measured a cytoplasmic electrical field of up to 0.5 V.cm^{-1}. In an elegant study Kropf (1986) measured a cytoplasmic electrical field of up to 0.4 V.cm^{-1} in hyphae of *A. bisexualis*. This field was abolished under conditions where no current flowed and hyphae did not extend. The size of these fields are of the same order as the minimal field strengths capable of inducing galvanotropism of hyphae of *N. crassa* (McGillivray and Gow, 1986). From consideration of the electrophoretic mobility of charged proteins and organelles (0.8 μm.s^{-1}.V^{-1}.cm and 4 μm.s^{-1}.V^{-1}.cm) (Sober, 1970; Jaffe, 1969; Sack *et al.*, 1983) and ignoring back diffusion, a field of up to 0.5 V.cm^{-1} should be capable of transporting these components apically at a rate of up to 25 μm.min^{-1} and 120 μm.min^{-1} respectively. These fields are on a theoretical basis more than large enough to deliver membrane vesicles and proteins to even the most rapidly extending hyphal tip. There is, however, good reason to believe that intracellular transport is predominantly the job of the actin and tubulin proteins of the cytoskeleton (see chapter by Hyams in this volume). We do not know whether the microsomal vesicles (such as chitosomes) which support tip growth are highly charged or whether cytosolic proteins in hyphae are distributed according to their charge. Some evidence does exist however, for the electrophoretic sorting of proteins in an electrically active tissue. Woodruff and Telfer (1973) demonstrated a 5−10 mV potential difference between the nurse cells and oocyte of *Crecropia* moth ovarioles so that the oocyte was relatively depolarized (more positive). Fluorescently labelled proteins of a known charge were microinjected into either nurse cells or the oocyte and the distribution of protein monitored by observing the spread of fluorescence (Woodruff and Telfer, 1980). Positively charged proteins localized in the nurse cells and negatively charged ones in the oocyte. If electropositive lysozyme was converted by carboxymethylation to an electronegative protein its direction of migration was reversed. Neutral proteins moved freely between the nurse cells and oocytes of the *Cecropia* follicle. Applied electric fields have been shown to induce changes in the actin microfilament conformation in *Xenopus* cells (Luther *et al.*, 1983) and may also generate electroosmotic flow within a cell (Jennings, 1986). Thus there are several ways in which electric fields may induce and polarize intracellular transport of cell components.

Electric fields may also influence the distribution of proteins in the cell membrane which bear exposed charged groups. Jaffe (1977) calculated that a voltage drop of only 0.8−4.0 mV across a cell could be enough to polarize (by between 10% and 50%) membranous particles or proteins of a diameter of 20 nm. The approximate time required to reach this degree of polarization is given by L^2/D, where L is the cells diameter and D is the mobile components diffusion constant. Applying this to a fungal spore of diameter 10 μm and assuming a diffusion constant of a membrane protein to be around 10^{-9} cm^{-2}.s^{-1} (Edidin, 1981), the time required for polarization would be in the order of 17 min. This is well within the response time for galvanotropic reactions of mycelial fungi (Gow, unpublished) and much shorter than the time over which natural currents flow. Moreover those endogenous currents and fields that predict fungal branch formation (Kropf *et al.*, 1983; Gow and Gooday, in preparation) often precede it by more than 15 min. Thus there is sufficient voltage and time for natural electrical fields to influence the distribution of membrane proteins in fungal hyphae.

A number of membrane proteins including the receptor for Concanavalin A (Con A) (Poo and Robinson, 1977), acetyl choline (Orida and Poo, 1978), and Fc_{ϵ} (immunoglobulin E receptor) (McCloskey *et al.*, 1984) have been shown to migrate to the cathodal side of cells that were exposed to physiological electrical fields (Poo, 1981). Glycoproteins like these often have a net negative charge at physiological pHs and so it might have been expected that such proteins would move to the anode. McLaughlin and Poo (1981) have therefore suggested that these proteins migrate not by direct electrophoresis but rather by electroosmosis in which counter ions to the proteins negative charges are electrophoresed to the cathode and in so doing set up a water flow which sweeps the exposed domains of the proteins in the same direction. This hypothesis is supported by the observation that Con A and acetylcholine receptors can be made to move to the anode if surface negative groups are removed using a neuraminidase enzyme (McLaughlin and Poo, 1981). Remarkable experiments by Patel and Poo (1982) showed that inhibition of the electric field induced redistribution of Con A receptors to the cathodic ends of *Xenopus* neurones (through the simple procedure of adding Con A), also abolished the capacity of these cells to exhibit galvanotropism. This strongly suggests that the field-induced redistribution of membrane proteins is part of the mechanism accounting for the galvanotropic behaviour of these cells. In fungal hyphae we know that chitin synthetase is not polarized by artificial electric fields (DeVries and Wessels, 1982) even though the activity of this enzyme is confined to the apex.

Ionic consequences

Electrical fields whether endogenous or applied, drive currents of ions through cells and lead to the establishment of cytoplasmic ion gradients. These gradients may arise as the natural consequence of local ion entry through the endogenous current carrying proteins or as a result of electric field induced alterations to the driving force for ion transport at either end of a cell. Gradients of protons and particularly calcium ions have been implicated in the control of the spatial organization of eukaryotic cells (Jaffe *et al.*, 1975; Robinson and McCaig, 1980). Proton gradients in fungi driving proton currents have been demonstrated using microelectrodes and pH-sensitive permeable dyes (Gow *et al.*, 1984; Turian, 1979; McGillivray and Gow, in press). A variety of techniques including use of low temperature ^{45}Ca autoradiography, calcium-sensitive microelectrodes, proton microprobes or fluorescent dyes such as chlorotetracycline, aequorin, quin 2 or fura 2 have revealed calcium gradients within living cells (Jaffe *et al.*, 1975; Reiss and Herth, 1979; Bosch *et al.*, 1980; Brownlee and Wood, 1986; Poenie and Tsien, 1986). In tip-growing organisms that have been studied calcium has been found to be locally at high concentration at the growing end. The cytoplasm in the hyphal tips of fungi such as *A. bisexualis* and *N. crassa* which circulate currents of protons is probably acidic and locally high in free calcium. That this is of significance in the establishment of polarity is evident from studies showing that proton and calcium ionophores induce branching in *A. bisexualis* (Harold and Harold, 1986). The calcium ionophore A23187 also stimulates branching in *N. crassa* (Reisig and Kinney, 1983) and has been used to polarize the outgrowth of zygotes of *Pelvetia* (Robinson and Cone,

1980) and to induce and polarize rhizoid formation in *Funaria* (Chen and Jaffe, 1979; Saunders and Hepler, 1982). Calcium gradients have also been implicated in the control of local growth and differentiation of the lobed alga *Micrasterias thomasiana* (McNally *et al.*, 1983). Proton ionophores may in fact also lead to changes in the cytosolic levels of calcium since they are known to cause calcium release from mitochondria. Calcium ions often have very low diffusion constants in cytoplasm compared to ions like K^+ or Na^+ due to the binding of calcium to fixed negative charges. Calcium gradients themselves can therefore generate fixed charge gradients (Donnan potentials) within cells which may be maintained even when the vectorial transcellular ion flux is small or even reversed. Electrical fields may therefore affect local intracellular calcium concentrations and local intracellular pH with consequences to polarity. We do not know how these local differences in the ionic environment might be translated into local growth. They may provide a permissive environment for vesicle fusion, enzyme activity or modulate the activities of the cytoskeleton. The link between calcium currents, the cytoskeleton and polarity is evident from the study of Brawley and Robinson (1985) who show that F-actin polymerization at the site of outgrowth of *Pelvetia* may be controlled by local calcium entry at this point.

Conclusions

Tip growth and branching of fungal hyphae is associated with the generation of endogenous ion currents and electric fields. Applied electric fields of a similar magnitude polarize hyphal growth and stimulate branching. Electrical fields are capable of localizing charged cellular components in the cytoplasm and cell membrane and establishing cytoplasmic gradients of ions such as protons and calcium. Some or all of these effects may be important to the transport of proteins or vesicles to the hyphal apex or to the selective accumulation and fusion of vesicles at this site. Ion currents and electrical fields may impinge directly on these processes or may act indirectly by regulating the activities of the cytoskeleton. However, we can still only speculate about the precise mechanisms that link ion fluxes, charge and ionic gradients to localized hyphal extension and branch formation.

References

Armbruster,B.L. and Weisenseel,M.H. (1983) Ionic currents traverse growing hyphae and sporangia of the mycelial water mould *Achlya debaryana*. *Protoplasma*, **115**, 665–669.

Bartnicki-Garcia,S., Ruiz-Herrera,J. and Bracker,C.E. (1979) Chitosomes and chitin synthesis. In *Fungal Walls and Hyphal Growth*. Burnett,J.H. and Trinci,A.P.J. (eds), Cambridge University Press, Cambridge, pp. 149–168.

Behrens,H.M., Weisenseel,M.H. and Sievers,A. (1982) Rapid changes in the pattern of electric current around the root tip of *Leipidium sativum* L. following gravistimulation. *Plant Physiology*, **70**, 1079–1083.

Bentrup,F.W. (1968) Die Morphogenese pflanzlicher Zellen im elektrischen Feld. *Zeitschrift für Pflanzenphysiologie*, **59**, 309–339.

Bertholon,M. (1783) *De l'électricité des vegetaux*. Paris.

Bosch,F., El Goresy,A., Herth,W., Martin,B., Nobiling,R., Povh,B., Reiss,H.D. and Traxel,K. (1980) The Heidelberg proton microprobe. *Nuclear Science Applications*, **1**, 33–55.

Brawley,S.H. and Robinson,K.R. (1985) Cytochalasin treatment disrupts the endogenous currents associated with cell polarisation in Fucoid zygotes: Studies of the role of f-actin in embryogenesis. *Journal of Cell Biology*, **100**, 1173–1184.

Brawley,S.H., Wetherell,D.F. and Robinson,K.R. (1984) Electrical polarity in embryos of wild carrot precedes

cotyledon differentiation. *Proceedings of the National Academy of Sciences USA*, **81**, 6064–6067.
Brower,D.L. and McIntosh,J.R. (1980) The effects of applied electric fields on *Micrasterias*. *Journal of Cell Science*, **42**, 261–277.
Brownlee,C. and Wood,J.W. (1986) A gradient of cytoplasmic free calcium in growing rhizoid cells of *Fucus serratus*. *Nature*, **320**, 624–626.
Chen,T.-H. and Jaffe,L.F. (1979) Forced calcium entry and polarized growth of *Funaria* spores. *Planta*, **144**, 401–406.
Cooper,M.S. and Keller,R.E. (1984) Perpendicular orientation and directional migration of amphibian neural crest cells in dc electric fields. *Proceedings of the National Academy of Science USA*, **81**, 160–164.
Cooper,M.S. and Schliwa,M. (1985) Electrical and ionic control of tissue cell locomotion in DC electric fields. *Journal of Neuroscience Research*, **13**, 223–244.
De Vries,S.C. and Wessels,J.G.H. (1982) Polarized outgrowth of hyphae by constant electrical fields during reversion of *Schizopyllum commune* protoplasts. *Experimental Mycology*, **6**, 95–98.
Dorn,A. and Weisenseel,M.H. (1982) Advances in vibrating probe techniques. *Protoplasma*, **113**, 89–96.
Edidin,M.A. (1981) Molecular motions and membrane organisation and function. In *Membrane Structure*. Finean,J.B. and Michell,R.H. (eds), Elsevier, Amsterdam, pp. 37–82.
Erickson,C.A. and Nuccitelli,R. (1982) Embryonic cell motility can be guided by weak electric fields. *Journal of Cell Biology*, **95**, (2 Part 2) 314a.
Erickson,C.A. and Nuccitelli,R. (1984) Embryonic fibroblast motility and orientation can be influenced by physiological electric fields. *Journal of Cell Biology*, **98**, 296–307.
Freeman,J.A., Manis,P.B., Snipes,G.J., Mayes,B.N., Samson,P.C., Wikswo,J.P. and Freeman,D.B. (1985) Steady growth cone currents revealed by a novel circularly vibrating probe. *Journal of Neuroscience Research*, **13**, 257–283.
Gooday,G.W. (1983) The hyphal tip. In *Fungal Differentiation: a Contemporary Synthesis*. Smith,J.E. (ed.), Marcel Dekker, New York, pp. 315–356.
Gow,N.A.R. (1984) Transhyphal electrical currents in fungi. *Journal of General Microbiology*, **130**, 3313–3318.
Gow,N.A.R. and Gooday,G.W. (1982) Effects of antheridiol on growth, branching and electrical currents of hyphae of *Achlya ambisexualis*. *Journal of General Microbiology*, in press.
Gow,N.A.R., Kropf,D.L. and Harold,F.M. (1984) Growing hyphae of *Achlya bisexualis* generate a longitudinal pH gradient in the surrounding medium. *Journal of General Microbiology*, **130**, 2967–2974.
Harold,R.L. and Harold,F.M. (1986) Ionophores and cytochalasins modulate branching in *Achlya bisexualis*. *Journal of General Microbiology*, **132**, 213–219.
Hinkle,L., McCaig,C.D. and Robinson,K.R.(1981) The direction of growth of differentiating neurones and myoblasts from frog embryos in an applied electric field. *Journal of Physiology (London)*, **314**, 121–135.
Horwitz,B.A., Weisenseel,M.H., Dorn,A. and Gressel,J. (1984) Electric currents around growing *Trichoderma* hyphae, before and after photoinduction of conidiation. *Plant Physiology*, **74**, 912–916.
Jaffe,L.F. (1966) Electrical currents through the developing *Fucus* egg. *Proceedings of the National Academy of Sciences USA*, **56**, 1102–1109.
Jaffe,L.F. (1969) On the centripetal course of development, the *Fucus* egg, and self electrophoresis. *Developmental Biology Supplement*, **3**, 83–111.
Jaffe,L.F. (1977) Electrophoresis along cell membranes. *Nature*, **265**, 600–602.
Jaffe,L.F. (1981) The role of ionic currents in establishing developmental pattern. *Philosophical Transactions of the Royal Society of London*, **B295**, 553–566.
Jaffe,L.F. and Nuccitelli,R. (1974) An ultrasensitive vibrating probe for measuring steady extracellular currents. *Journal of Cell Biology*, **63**, 614–628.
Jaffe,L.F. and Poo,M.-M. (1979) Neurites grow faster towards the cathode than the anode in a steady field. *Journal of Experimental Zoology*, **209**, 115–120.
Jaffe,L.F., Robinson,K.R. and Nuccitelli,R. (1974) Local cation entry and self-electrophoresis as an intracellular localisation mechanism. *Annals of the New York Academy of Sciences USA*, **238**, 372–389.
Jaffe,L.A., Weisenseel,M.H. and Jaffe,L.F. (1975) Calcium accumulations within the growing tips of pollen tubes. *Journal of Cell Biology*, **67**, 488–492.
Jennings,D.H. (1986) Morphological plasticity in fungi. In *Plasticity in Plants*. Jennings,D.H. and Trewavas,A.J. (eds), Symposia of the Society for Experimental Biology No. XXXX. The Company of Biologists, Cambridge, pp. 329–346.
Khew,K.L. and Zentmeyer,O.A. (1971) The electrotactic response of zoospores of seven species of *Phytophthora* fungi. *Phytopathology*, **64**, 500–507.
Kropf,D.L. (1986) Electrophysiological properties of *Achlya* hyphae: ionic currents studied by intracellular

potential recording. *Journal of Cell Biology*, **102**, 1209–1216.
Kropf,D.L., Caldwell,J.C., Gow,N.A.R. and Harold,F.M. (1984) Transcellular ion currents in the water mold *Achlya*. Amino acid proton symport as a mechanism of current entry. *Journal of Cell Biology*, **99**, 486–496.
Kropf,D.L., Lupa,M.D., Caldwell,J.C. and Harold,F.M. (1983) Cell polarity: endogenous ion currents precede and predict branching in the water mold *Achlya. Science*, **220**, 1385–1387.
Lund,E.J. (1923) Electrical control of organic polarity in the egg of *Fucus. Botanical Gazette*, **76**, 288–301.
Lund,E.J. (1947) *Bioelectric Fields and Growth*. University of Texas Press, Austin, Texas, USA.
Luther,P.W., Peng,H.B. and Lin,J.J.-C. (1983) Changes in cell shape and actin distribution induced by constant electric fields. *Nature*, **303**, 61–64.
Marsh,G. and Beams,H.W. (1945) The orientation of pollen tubes of *Vinca* in the electric current. *Journal of Cellular and Comparative Physiology*, **25**, 195–204.
McCaig,C.D. (1986) Dynamic aspects of amphibian neurite growth and the effects of an applied electric field. *Journal of Physiology*, **375**, 55–59.
McCloskey,M.A., Liu,Z.-Y. and Poo,M.-M. (1984) Lateral electromigration and diffusion of Fc_ϵ receptors on rat basophilic leukemia cells: effects of IgE binding. *Journal of Cell Biology*, **99**, 778–787.
McLaughlin,S. and Poo,M.-M. (1981) The role of electro-osmosis in the electric field-induced movement of charged macromolecules on the surfaces of cells. *Biophysical Journal*, **34**, 85–94.
McGillivray,A.M. and Gow,N.A.R. (1986) Applied electrical fields polarize the growth of mycelial fungi. *Journal of General Microbiology*, **132**, 2515–2525.
McGillivray,A.M. and Gow,N.A.R. (1987) The transhyphal electrical current of *Neurospora crassa* is carried principally by protons. *Journal of General Microbiology*, in press.
McNally,J.G., Cowan,J.D. and Swift,H. (1983) The effects of the ionophore A23187 on pattern formation in the alga *Micrasterias. Developmental Biology*, **97**, 137–145.
Miller,A.L., Raven,J.A., Sprent,J.I. and Weisenseel,M.H. (1986) Endogenous ion currents traverse growing roots and root hairs of *Trifolium repens. Plant, Cell and Environment*, **9**, 79–83.
Müller,W.A. and El-Shershaby,E. (1981) Electrical current and cAMP induce lateral branching in the stolon of hydroids. *Developmental Biology*, **87**, 24–29.
Nawata,T. (1984) A simple method for making a vibrating probe system. *Plant and Cell Physiology*, **25**, 1089–1094.
Nuccitelli,R. (1978) Ooplasmic segregation and secretion in the *Pelvetia* egg is accompanied by a membrane generated electrical current. *Developmental Biology*, **62**, 13–33.
Nuccitelli,R. (1986) A two-dimensional vibrating probe with computerized graphics display. In *Ionic Currents in Development*. Nuccitelli,R. (ed.), Alan R.Liss, New York, pp. 13–20.
Nuccitelli,R. and Erickson,C.A. (1983) Embryonic cell motility can be guided by physiological electric fields. *Experimental Cell Research*, **147**, 195–201.
Nuccitelli,R. and Jaffe,L.F. (1974) Spontaneous current pulses through developing fucoid eggs. *Proceedings of the National Academy of Sciences USA*, **71**, 4855–4859.
Onuma,E.K. and Hui,S.W. (1985) A calcium requirement for electric field-induced cell shape changes and preferential orientation. *Cell Calcium*, **6**, 281–292.
Orida,N. and Poo,M.-M. (1978) Electrophoretic movement and localisation of acetyl-choline receptors in the embryonic muscle cell membrane. *Nature*, **275**, 31–35.
Patel,N.B. and Poo,M.-M. (1982) Orientation of neurite growth by extracellular electric fields. *Journal of Neurosciences*, **2**, 483–496.
Patel,N.B. and Poo,M.-M. (1984) Perturbation of the direction of neurite growh by pulsed and focal electric fields. *Journal of Neurosciences*, **4**, 2939–2947.
Peng,H.B. and Jaffe,L.F. (1976) Polarization of Fucoid eggs by steady electrical fields. *Developmental Biology*, **53**, 277–284.
Poenie,M. and Tsien,R. (1986) Fura-2: A powerful new tool for measuring and imaging $[Ca^{2+}]_i$ in single cells. In *Ionic Currents in Development*. Nuccitelli,R. (ed.), Alan R.Liss, New York, pp. 53–56.
Poo,M.-M. (1981) In situ electrophoresis of membrane components. *Annual Reviews of Biophysics and Bioengineering*, **10**, 245–276.
Poo,M.-M. and Robinson,K.R. (1977) Electrophoresis of concanavalin A receptors along embryonic muscle cell membrane. *Nature*, **265**, 602–605.
Rathore,K.S. (1986) Electrical stimulation of growth and differentiation in plant tissue cultures. In *Ionic Currents in Development*. Nuccitelli,R. (ed.), Alan R.Liss, New York, pp. 295–302.
Rathore,K.S. and Goldsworthy,A. (1985) Electrical control of growth in plant tissue cultures. *Biotechnology*, **3**, 253–254.

Reiss,H.-D. and Herth,W. (1979) Calcium gradients in tip growing plant cells visualised by chlorotetracycline fluorescence. *Planta*, **146**, 615–621.

Reissig,J.L. and Kinney,S.G. (1983) Calcium as a branching signal in *Neurospora. Journal of Bacteriology*, **154**, 1392–1402.

Robinson,K.R. (1985) The responses of cells in electric fields: a review. *Journal of Cell Biology*, **101**, 2023–2027.

Robinson,K.R. and Cone,R. (1980) Polarization of fucoid eggs by calcium ionophore gradient. *Science*, **207**, 77–78.

Robinson,K.R. and Jaffe,L.F. (1975) Polarizing fucoid eggs drive a calcium current through themselves. *Science*, **187**, 70–72.

Robinson,K.R. and McCaig,C.D. (1980) Electrical fields, calcium gradients, and cell growth. *Annals of the New York Academy of Sciences*, **339**, 132–138.

Sack,F.D., Priestley,D.A. and Leopold,A.C. (1983) Surface charge on maize-coleoptile amyloplasts. *Planta*, **157**, 511–517.

Sand,O. (1973) On orientation of rhizoid outgrowth of *Ulva mutabilis* by applied electric fields. *Experimental Cell Research*, **76**, 444–446.

Saunders,M.J. and Hepler,P.K. (1982) Calcium ionophore A23187 stimulates cytokinin-like mitosis in *Funaria. Science*, **217**, 943–945.

Schrank,A.R. (1947) Electrical and curvature responses of the *Avena* coleoptile to transversely applied direct current. In *Bioelectric Fields and Growth.* Lund,E.J. (ed.), Texas University Press, Austin, Texas, USA, pp. 217–231.

Slayman,C.L. and Slayman,C.W. (1962) Measurements of membrane potential in *Neurospora. Science*, **136**, 876–877.

Sober,H.A. (1970) In *Handbook of Biochemistry.* Second edition, Sober,H.A. (ed.), CRC Press, Cleveland, pp. C36.

Stump,R.F. and Robinson,K.R. (1983) *Xenopus* neural crest cell migration in an applied electric field. *Journal of Cell Biology*, **97**, 1226–1233.

Stump,R.F., Robinson,K.R., Harold,R.L. and Harold,F.M. (1980) Endogenous electrical currents in the water mold *Blastocladiella emersonii* during growth and sporulation. *Proceedings of the National Academy of Sciences USA*, **77**, 6673–6677.

Troutman,J.I. and Wills,W.H. (1964) Electrotaxis of *Phytophthora parasitica* zoospores and its possible role in infection of tobacco by the fungus. *Phytopathology*, **54**, 225–228.

Turian,G. (1979) Cytochemical gradients and mitochondrial exclusion in the apices of vegetative hyphae. *Experientia*, **37**, 1164–1166.

Waaland,S.D. and Lucas,W.J. (1984) An investigation of the role of transcellular ion currents in morphogenesis of *Griffithsia pacifica* Kylin. *Protoplasma*, **123**, 184–191.

Weisenseel,M.H. and Jaffe,L.F. (1976) The major growth current through lily pollen tubes enters as K^+ and leaves as H^+. *Planta*, **133**, 1–7.

Weisenseel,M.H., Dorn,A. and Jaffe,L.F. (1979) Natural H^+ currents traverse growing roots of barley (*Hordeum vulgare* L.). *Plant Physiology*, **64**, 512–518.

Weisenseel,M.H., Nuccitelli,R. and Jaffe,L.F. (1975) Large electrical currents traverse growing pollen tubes. *Journal of Cell Biology*, **66**, 556–567.

Woodruff,R.I. and Telfer,W.H. (1973) Polarized intercellular bridges in ovarian follicles of the *Cecropia* moth. *Journal of Cell Biology*, **58**, 172–178.

Woodruff,R.I. and Telfer,W.H. (1980) Electrophoresis of proteins in intercellular bridges. *Nature*, **286**, 84–86.

CHAPTER 4

The partitioning of cytoplasmic organelles at cell division

C.WILLIAM BIRKY, Jr

Department of Molecular Genetics, The Ohio State University, 484 West 12th Avenue, Columbus, OH 43210 USA

The problems

When a cell divides, its components must be divided up or partitioned between the daughter cells. This chapter focuses on the partitioning of a particular class of organelles, those which are known to contain DNA and genes and to be self-replicating. At present, this class includes only mitochondria and chloroplasts, together with intracellular symbionts that are essential for their host. The mechanism(s) involved in their partitioning is our subject matter. It is common to speak about biological mechanisms as if they were solutions to a series of problems faced and solved by organisms. I will do so here, remembering of course that these problems are solved, not by the volition of the organisms, but by evolutionary processes that mobilize the physical and chemical mechanisms. Cell division poses two general kinds of partitioning problems for self-replicating organelles and their DNA.

(i) When a cell divides, each daughter cell must receive at least one copy of each organelle and at least one complete copy of each organelle genome. The cell biologist sees this as being necessary to ensure the survival of the daughter cells. The geneticist sees it as necessary to ensure hereditary continuity. A daughter cell with no chloroplast or mitochondrion cannot make one *de novo* and cannot live without it. In general, the daughter cell must also receive copies of the chloroplast or mitochondrial DNA molecules as well, for the organelles will not be fully functional without them. These are very strong generalizations, but I know of no eukaryote which normally has mitochondria but which can live and reproduce in nature without them. A few organisms, such as the yeast *Saccharomyces cerevisiae*, are viable without functional mitochondrial DNA so long as they are fed fermentable substrates, but they still require mitochondria and are generally weak relative to their normal counterparts. Some algae and plants that are facultative heterotrophs can exist without functional genes for photosynthesis. But all of these that have been carefully studied have been found to retain plastids (leucoplasts) and plastid DNA of unknown but essential function (e.g. Hussein *et al.*, 1982).

(ii) It is generally not sufficient for a cell to have just one copy of an organelle DNA molecule, or some arbitrarily small portion of a mitochondrion or chloroplast. It must have the correct amount of organelle activity, organelle membrane area, amount of photosynthetic or respiratory enzymes, number of DNA molecules and ribosomes, and so forth. This can be ensured by controlling organelle partitioning, or growth and replication, or both.

These same problems are faced by another self-replicating organelle, the nuclear chromosome. It is solved, first, by controlling chromosome replication so that each chromosome replicates precisely once in each cell cycle; second, by mitosis which results in equal partitioning, so that each daughter cell gets one copy of each chromosome. Chromosome replication and partitioning are both under stringent control, meaning that every cell replicates and partitions its chromosomes the same way in every cell division. That is not to say that the control is perfect. For example, mistakes are made in partitioning so that one daughter cell receives two copies of a chromosome and the other daughter cell receives none. But these mistakes are so rare that we give them a special name, nondisjunction. No mechanism has evolved to correct for them, presumably because developing and maintaining the mechanism would be more expensive for a species than is the occasional loss of a cell or individual. In contrast, there are some well-known examples of microorganisms in which nuclear division is amitotic. In the case of the ciliate macronucleus, this results in daughter nuclei and cells that receive variably unequal amounts of chromosomal DNA. This variability in partitioning is compensated by controlling DNA replication (Doerder, 1979; Berger, 1979).

In contrast to nuclei, there is no evident mitotic apparatus for cytoplasmic organelles. Since they cannot make use of the mitotic apparatus, other mechanisms have evolved to promote more or less equal partitioning of mitochondria and chloroplasts; the first part of this chapter is a review of these mechanisms as they are seen at the cytological level. The most important conclusion is that evolution has been opportunistic in solving this problem, using a variety of different cellular mechanisms even in closely related organisms. In the second part of this chapter I will review the rather meagre quantitative data, which will show that partitioning mechanisms for organelles are imperfect. In other words, the partitioning of organelles is under relaxed control, so that daughter cells receive variable and unequal numbers or mass of organelles. Nevertheless, the partitioning of organelles is more equal than would be expected if it were left completely to chance. This is probably also true for organelle DNA molecules, but there are no relevant quantitative data. The third section of the chapter reconsiders the mechanisms of partitioning, asking to what extent the non-randomness of organelle partitioning can be explained in a gross mechanical way, by volume effects or by the molecular properties of the cytoskeleton.

Patterns of organelle partitioning

For the sake of organization, it is convenient to distinguish between four general patterns of organelle partitioning, which can be thought of as four different kinds of solutions

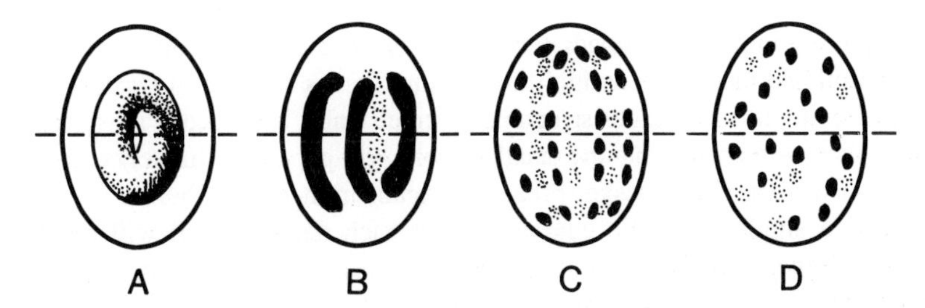

Figure 1. Four general patterns of organelle partitioning. The organelles are shaded; the plane of division is indicated by the dashed lines. **A**, one organelle in a fixed position such that it divides in half before or during cytokinesis and each daughter receives one half; **B**, multiple organelles, each likewise divided in half; **C**, multiple organelles in fixed positions such that each daughter receives one half of the number; **D**, multiple organelles arranged and partitioned stochastically.

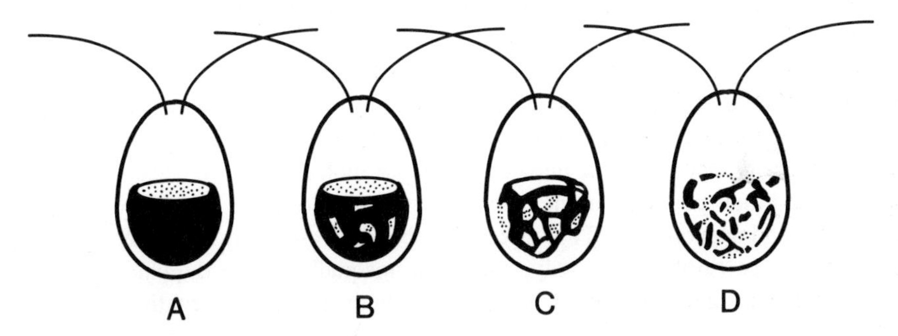

Figure 2. The continuum from a single large organelle (here, cup-shaped like the chloroplast of *Chlamydomonas*) with varying degrees of fenestration, to many small organelles.

to the partitioning problems. These are shown diagrammatically in *Figure 1*; specific examples are cited in Birky (1983).

One organelle cut in half

Many microorganisms have a single chloroplast, which is located in a fixed, symmetrical position in the cell, such that it is bisected by the cleavage plane. The chloroplast divides in the cleavage plane, just before or coincident with cytokinesis. The daughter cells will then receive approximately equal volumes of organelle, mass of organelle membranes, and amounts of enzymes and other molecules. Some species have a single mitochondrion which is partitioned in this manner, although it appears to be less common. A well-known example is the cup-shaped chloroplast of *Chlamydomonas reinhardtii* which is cut approximately in half in the vertical plane by cytokinesis. In this case the organelle is very regular in shape, but it may also be an irregular basketwork or a set of interconnected filaments, as in the leucoplast of *Polytoma*. In fact, if one looks at different organisms and organelles, one can find a continuum of cases with no sharp dividing lines between the regular cup with or without fenestrations, baskets with more or less open weave, and individual organelles, as illustrated in *Figure 2*. Moreover, a single organelle can undergo transitions between these forms. This is seen, for example, in the mitochondrion of *Polytoma*, which is seen as an open basket in late interphase; sometime immediately before or after cell division the basket breaks up into individual organelles, which fuse again. This is significant because the fusion of organelles to form a single structure symmetrically arranged with respect to the cleavage plane will ensure that the daughter cells will get approximately equal volumes of organelles, even if the basket breaks up again immediately before cytokinesis.

Bisecting a single organelle guarantees that each daughter cell will receive some organelle volume, roughly half of the parental organelle, but it does not ensure precisely equal partitioning. I have seen no quantitative data on the extent to which daughter cells receive different portions of the parental organelle. Volume measurements could be done on serial electron micrographs of dividing cells, but there seems to be no compelling reason to invest the immense labour of doing this on a statistically significant sample of cells. It would be relatively easy to measure chlorophyll or other molecules in newly divided cells by fluorescence microspectrophotometry.

This mode of partitioning of the organelle volume does not necessarily contribute to equal partitioning of the organelle DNA molecules. Fluorescent staining of the chloroplast DNA (cpDNA) molecules in a haploid *Chlamydomonas* shows that they are aggregated into a variable number of nucleoids averaging about five; the size and staining intensity, and hence the number of cpDNA molecules in the nucleoids, is highly variable, and their positions in the chloroplast show no obvious regularity (e.g. Birky *et al.*, 1984 and unpublished). Thus there is no obvious reason why the cpDNA molecules would not be partitioned randomly, and often unequally, while the chloroplast itself is partitioned roughly equally. Unfortunately there are no quantitative data on the partitioning of mitochondrial DNA (mtDNA) or cpDNA molecules between daughter organelles or daughter cells. In a number of studies, mtDNA or cpDNA has been labelled with [^{3}H]thymidine and grain counts made on autoradiographs of daughter organelles or cells. Unfortunately, although this technique shows that there is no tendency to give consistently more DNA molecules to one daughter than the other, it cannot distinguish between the randomness inherent in the technique and any possible randomness in partitioning of the molecules (Birky, 1983).

On the other hand, there are cases in which organelles contain a single nucleoid which lies symmetrically to the plane of division of the organelle. Consequently, organelle division divides these nucleoids (and presumably their content of DNA molecules) roughly in half, just as cell division divides a single organelle in half. A striking example is the cigar-shaped nucleoid seen in the mitochondrion of the slime mold *Physarum polycephalum*, which is cut in half when the mitochondrion divides (Kuroiwa *et al.*, 1977). Presumably this means that the daughter mitochondria each get about half of the mtDNA molecules.

Multiple organelles, each divided in half

A varient on the preceding theme involves multiple organelles, each positioned so that it is bisected by the cleavage plane and each divided in half during cytokinesis. This is well known in the mitochondria of some insects and scorpions (reviewed by Birky, 1983). A striking example in the protists is the green alga *Spirogyra*, in which the helically arranged chloroplasts in a cell are divided transversely at cytokinesis.

Multiple organelles partitioned uniformly

In some cases there are multiple organelles located in fixed positions with equal numbers on the two sides of the cleavage plane. Then cytokinesis automatically partitions the organelles equally by number to the daughter cells. In this case, organelle division and cytokinesis may occur at different times. The partitioning may not be precisely equal in some cases, where the cell division is asymmetrical to produce differentiated daughter cells. But if there is a mechanism for positioning the organelles at specific sites, their partitioning will be very uniform from division to division. The cortical mitochondria of *Tetrahymena thermophila* are a striking example, carefully analysed by Aufderheide (1979). The cortex of this ciliated protozoan is composed of repeating units, each of which contains the basal body of a cilium, a regular arrangement of microtubules, and a mitochondrion. These units are arranged in longitudinal rows called kineties. As the cell grows, each kinety grows longitudinally by duplicating each unit. When the cell divides transversely, each daughter cell receives approximately half of each kinety, and thus half of the cortical mitochondria. These cells also contain mitochondria in the fluid endoplasm, which are continually exchanged with the cortical mitochondria. Thus the cortex serves as a mechanism for partitioning a part of the total mitochondrial population equally between the daughter cells. There is probably some inequality between anterior and posterior daughters, and some variability among different dividing cells, but it is clearly small. The remaining mitochondria, in the endoplasm, are presumably partitioned more or less randomly.

Multiple organelles partitioned stochastically

Some microbial cells contain many mitchondria or chloroplasts that appear to be distributed more or less randomly throughout the cytoplasm. These may be rather uniformly round, oval, or cigar-shaped as in the ascomycete fungus *Neurospora* and the slime mold *Physarum*. Alternatively they may be highly branched, in which case it might be more appropriate to say that the mitochondrial mass is distributed randomly throughout the cytoplasm. In the yeast *S. cerevisiae*, one may find both conditions depending upon the strain and the physiological state of the cell (Stevens, 1977). In some cases, the organelles are not distributed throughout the cytoplasm but are limited to a region which is bisected by cytokinesis. For example, the chloroplasts are limited to the periphery of the cell in the alga *Olisthodiscus luteus* which will be discussed in more detail below. But the arrangement of the organelles in this shell is clearly not regular and uniform as in the case of the cortical mitochondria of *Tetrahymena*; rather, it appears to be somewhat random.

In these cases, there are usually enough organelles in a cell to ensure that even if they were partitioned strictly randomly, the probability that a daughter cell will receive no organelles by chance is very small. But it appears that partitioning is not strictly random in these cases, being equal more often than expected by chance (as will be described below). In other words, the arrangement of the organelles is apparently more uniform than random, though by no means so uniform as is suggested by *Figure 1C*. An interesting variation on this theme is seen in *Pelomyxa palustris*, in which the symbiotic bacteria are normally dispersed through the cytoplasm but become attached

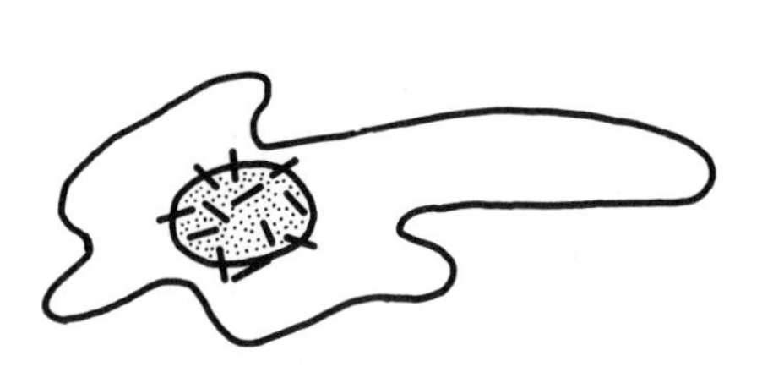

Figure 3. Symbiotic bacteria in the giant amoeba *Pelomyxa palustris* become associated with the nuclear membrane just before mitosis (**left**), so that when the nucleus divides amitotically, each daughter nucleus (and hence each daughter cell) receives some bacteria (**right**).

to the nuclear membrane just before mitosis (Whatley, 1976). Then, if they are distributed with some degree of uniformity over the nuclear membrane, they will automatically be partitioned so that each daughter cell will receive roughly half of the symbionts because it receives one daughter nucleus (*Figure 3*).

The evolution of partitioning mechanisms

The diversity of mechanisms used for cytoplasmic partitioning is in striking contrast to the uniformity of mechanisms used for nuclear genes. Most eukaryotes have solved the partitioning problem for chromosomes in the same way, by using a microtubular mitotic apparatus. There are some notable variations on the theme, but these are limited to specific taxa. For example, the dinoflagellates do not use a mitotic spindle of the usual sort, and fungi do the partitioning within the intact nuclear membrane, although they have microtubular spindles of a fairly conventional sort. In contrast, cytoplasmic organelles are partitioned in many different ways, and these show no obvious phylogenetic pattern. If mitochondria and chloroplasts arose by endosymbiosis, as is now almost certain, then a partitioning mechanism had to be developed at each symbiotic event. There is no evidence that mitochondria arose more than once, while chloroplasts may have arisen as many as six times, separately in lineages leading to green algae, red algae, euglenoids, dinoflagellates, cryptomonads, and other chromophytes (reviewed by Whatley, 1983). One might expect these different groups to have evolved different mechanisms, but there is no apparent correlation; in fact one group, the green algae, shows all of the above patterns of chloroplast partitioning.

It appears that each solution to the partitioning problem evolved independently in most of the major taxa, and probably several times in some taxa. It is probable that partitioning mechanisms changed from time to time in various lineages, as the morphology and physiology of organisms changed. Evolution has been opportunistic, using whatever means is at hand for partitioning. We will not be able to speak of *the* partitioning of organelles, but will always have to specify which kind of organism and which kind of organelle is under discussion. And just as there are many different general patterns of partitioning at the cytological level, there will not be a single molecular or cellular mechanism.

How variable is partitioning?

We have considered various ways in which cells partition their mitochondria and chloroplasts. Each mode of partitioning was viewed as a solution to the problems of providing each daughter cell with at least some organelle mass and at least some copies of the organelle genome, preferably half of what was in the parent. How successful are these soulutions? To what extent is partitioning equal and uniform, as opposed to random? There are three aspects of partitioning that could be measured.

Organelle mass or activity

One could measure the total organelle volume, mass, enzyme activity, or other measure of activity in pairs of newly formed daughter cells. For example, one could measure organelle volume by electron microscope serial sections. This is extremely laborious and has never been done with enough pairs of cells to give quantitative information about the amount of variability in the partitioning. It would be possible to measure some individual components of organelles using quantitative microscopy; for example, one might measure chlorophyll autofluorescence with a microspectrophotometer. Most of these techniques have a considerable amount of unavoidable measurement error; the variance due to this source would have to be carefully separated from whatever variance is intrinsic to the partitioning process.

Numbers of organelle DNA molecules

Partitioning could be quantitated by estimating the numbers of organelle DNA molecules in many pairs of newly formed daughter cells. This could be done by microspectrophotometry with a fluorescent dye, with the same caveat as above. There are several studies in which parent cells were labelled with [^{3}H]thymidine and daughter cells autoradiographed to estimate their contents of mtDNA (reviewed by Birky, 1983). Unfortunately, no method has been developed to separate the variance in grain counts due to random partitioning from that which is due to the randomness inherent in the labelling and radioactive decay processes.

Numbers of organelles

This method can be used to measure the variability of partitioning in any organism which has a modest number of large, discrete organelles immediately before cell division. Moreover, those organelles must not divide coincidentally with cytokinesis. Then it is possible to count organelles in the two halves of cells which are in the process of division, or in the daughter cells shortly after division. This method will not measure the partitioning of organelle mass or enzyme activity, or of the organelle DNA molecules, because the size and DNA content of the organelles varies. Nevertheless, the numerical data which are provided can be almost completely free of error. Moreover, they are directly related to a problem of replication control which will be considered later. Another advantage of numerical data of this kind is that they can be compared to two extreme hypotheses.

(i) The one extreme hypothesis might be called the null hypothesis; it is that the cell exerts completely relaxed control, namely, no control at all, over partitioning. Then partitioning is *strictly random*, in the sense that each organelle has a constant probability p (usually 1/2) of entering a given daughter cell, regardless of the behaviour of any of the other organelles. Another way of looking at this hypothesis is that the partitioning of each organelle is independent of the others. Mathematically, the number or proportion x of organelles received by a daughter cell is a random variable, and the probability that a cell will receive a particular number of organelles is given by the binomial distribution.

(ii) The extreme opposite hypothesis is that the cell exerts such stringent control over partitioning that it is precisely the same in every cell division. I will call this *uniform partitioning*. In most cases the two daughter cells will not be differentiated from each other, and the hypothesis states that each daughter will receive precisely half of the organelles in the mother, in every case. (Obviously if the mother has an odd number of organelles equal to $2n + 1$, the equivalent to equal partitioning is that one daughter receives n organelles and the other, $n + 1$.) If the daughter cells were differentiated so that one consistently received some proportion x of the organelles in the mother cell and the other received the remaining proportion $(1 - x)$, the hypothesis would say that x is constant. The model for such stringent control is the mitotic or meiotic partitioning of chromosomes.

(iii) In between these extremes, there is a continuum of models which I will call *stochastic* partitioning. In a stochastic model, one can assign probabilities to various outcomes but can never predict an outcome. But equal or nearly equal partitioning occurs more often than expected on a binomial distribution.

Most of the data have been reviewed before (Birky, 1983; Hennis and Birky, 1984), and three of the five studies have dealt with plants or animals rather than microorganisms, so I will consider them only briefly. In no case was partitioning uniform. In one of the studies on plants (Anton-Lamprecht, 1967), the partitioning of mitochondria and chloroplasts was not significantly different from the strictly random model (Hennis and Birky, 1984). However, only 14 cells were examined and the data would not exclude a stochastic model. The single study on animals pertains to the partitioning of mitochondria during meiosis in male scorpions (Hood *et al.*, 1972; Wilson, 1925, 1931); partitioning was stochastic (Birky, 1983). Stochastic partitioning was also observed for the partitioning of chloroplasts in clover (Butterfass, 1969) and in the alga *Olisthodiscus* (Hennis and Birky, 1984). As an example of stochastic partitioning, in the latter study we observed partitioning in nine cells that had 24 chloroplasts each; in six cases, each daughter received 12 chloroplasts, and in three cases one received 11 and the other 13. The binomial distribution (strictly random partitioning) predicts fewer than three cases such as we observed, and more than six cases of more unequal partitioning. This kind of deviation was found consistently for mother cells containing from eight to 34 chloroplasts.

It is widely (but not universally) believed that mitochondria and chloroplasts originated as prokaryotic intracellular symbionts. If so, the absence of stringent control over their partitioning can be understood as an evolutionary relic from the days when the symbionts/organelles were newcomers under no cellular control whatsoever (Birky, 1983). Moreover, it might be instructive to study the partitioning of known symbionts. One of these is the cyanelle of *Cyanophora paradoxa*, which has a genome very much like that of a chloroplast and serves as a chloroplast for its host, but is a cyanobacterium in morphology and some aspects of biochemistry (reviewed by Wasmann *et al.*, 1987). The division of cyanelles and host cells is not synchronous and in fact occurs at different rates in different culture conditions, so that the number of cyanelles per cell varies from one to nine and the mean number varies with the culture conditions (Trench *et al.*, 1978). Hall and Claus (1963) state that partitioning is not always equal, under conditions where most cells have four to six cyanelles, but there are no published quantitative data.

When I began studying this organism, my culture contained many cells with more than two cyanelles. A preliminary examination of dividing cells showed that partitioning was usually equal, but a few cases of unequal partitioning were seen. When I began collecting quantitative data, the frequency of dividing cells had decreased as had the number of cyanelles per cell. Under the present culture conditions, cells that have not begun dividing have one or two cyanelles, which are often in the process of division. The diameter of the spherical symbionts is large compared to that of the elongated host cells, so that when a cell has two cyanelles, they lie one behind the other, and if a cyanelle is dividing, the daughter cells also lie one behind the other. When division has begun, as evidenced by the appearance of a cleavage furrow at the end of the cell, the cell becomes somewhat broader. Its two cyanelles, or one cyanelle with two daughters, gradually rotate so as to lie side-by-side, where they can be separated by the cell's longitudinal cleavage furrow. To date, 44 dividing cells have been counted in which the furrow had progressed past the midpoint of the cell where the symbionts lie, and partitioning was equal in all of them. Of these, there were three cases of 2:2 partitioning, two of 2:1 partitioning, and 39 cases of 1:1 partitioning. Only one cell has been seen to begin division while it had only a single cyanelle, and that cyanelle was in the process of division. As further evidence that unequal partitioning is rare, of the many hundreds of non-dividing cells that have been examined, all had at least one cyanelle. It is clear that partitioning is usually, if not always, equal under conditions where *Cyanophora* contains only one or two cyanelles per cell, but my earlier observations and those of Hall and Claus (1963) suggest that unequal partitioning may be more common when cells have more symbionts.

It is striking that perfectly uniform partitioning has never been observed, and that strictly random partitioning has been excluded whenever a large number of cells has been observed. One is tempted to conclude that partitioning is usually stochastic. This generalization remains weak until more organisms are studied, but in the meantime I will assume that it is widely applicable and proceed from there.

Mechanisms of non-random partitioning for multiple organelles

From the standpoint of cellular and molecular mechanisms, we could say that cells exert some degree of control over partitioning, but that the control is imperfect. In other words, cells have a mechanism, or more likely various mechanisms, for partitioning organelles equally, but the mechanisms do not work very well. This is not to say that

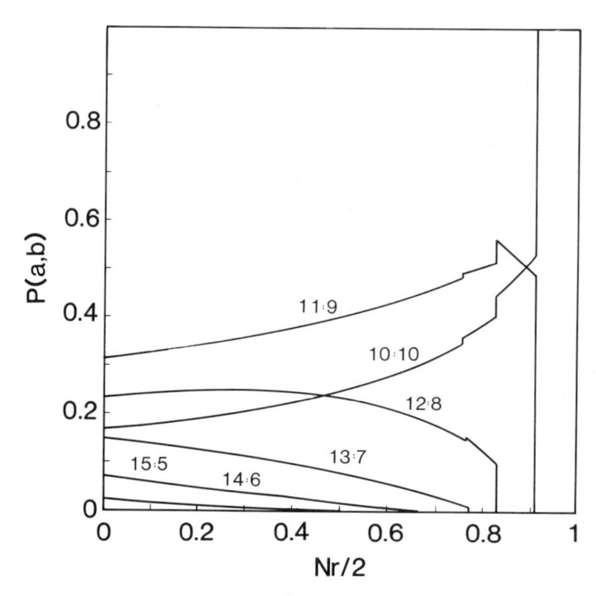

Figure 4. Volume effects on partitioning when the mother cell has $N = 20$ organelles. Each curve shows the probability $P(a,b)$ of a particular partitioning a:b (for example, 11:9 means one daughter cell receives 11 and the other 9 organelles), as a function of the ratio of total organelle volume to mother cell volume ($Nr/2$, where r is the ratio of the volume of one organelle to that of one daughter cell).

equal partitioning is always so advantageous that mechanisms have evolved specifically to promote it. When cells have many copies of an organelle, there may be very little cost associated with unequal partitioning, and hence little selective advantage to improving partitioning control. It is quite possible that stochastic partitioning has not evolved for its own sake, but is an indirect consequence of some very general property of cells or organelles.

Volume effects

One possibility is that the volume of the organelles may tend to force equal partitioning. Birky and Skavaril (1984) discussed this possibility and developed the relevant mathematical theory, only later discovering that the idea was first suggested by Butterfass (1979; Hennis and Birky, 1984). Briefly, the idea is that the binomial distribution, used for calculating the probabilities of various organelle distributions on the random model, assumes that the probability that a given organelle is included in a particular daughter cell is independent of the presence of any other organelles. This is a reasonable model only if the organelles are dimensionless points, that is if they have zero volume. Since each organelle has some volume, it occupies some space and thereby excludes other organelles from that space; hence organelles cannot be partitioned independently of each other. This will tend to increase the frequency of equal or nearly equal partitioning. To visualize this, imagine two basketballs and two small wastebaskets. If the first ball lands in the left-hand basket, the other ball must go in the right-hand basket because the left-hand basket is full.

Birky and Skavaril (1984) developed a modified binomial distribution which takes into account the number of organelles and the ratio of organelle volume to cell volume. *Figure 4* illustrates the volume effect for one specific case, where there are $N = 20$ organelles in the mother cell. Each organelle occupies a fraction r of the volume of a daughter cell, so that collectively the organelles occupy a fraction $Nr/2$ of the mother cell. The probability of an equal or nearly equal partitioning (10:10 or 11:9) increases slowly with increasing organelle volume. When the organelles occupy more than 90% of the cell volume, the only possible distribution is equal, since 11 organelles will not fit in one daughter cell. The important lesson of the mathematical theory is that volume effects are small unless the organelles occupy a very large portion of the cell volume; the available data (reviewed by Birky and Skavaril, 1984) are meagre but sufficient to show that the ratio of organelle to cell volume is rarely large enough.

On further reflection, this does not necessarily mean that volume effects cannot explain the observed non-randomness of partitioning. First, we noted that there are portions of the cell from which the organelles are excluded, notably the region occupied by other organelles and by the nucleus. Actually, we do not know if the relevant volume would be that of the nucleus or of the mitotic apparatus. Although in some organisms the nucleus is re-formed before cytokinesis, the mitotic apparatus may previously have forced organelles into a more or less uniform distribution. In any event, what is important for partitioning is not the total cell volume but the volume available to the organelles. I know of no case where the volume of a dividing cell, the mitochondria or chloroplasts, and the mitotic apparatus have been measured simultaneously. Even such data might not be sufficient, since there may be other fixed areas of the cell from which the organelles are excluded. This is suggested by electron micrographs of *Olisthodiscus* which show the chloroplasts apparently limited to an area near the periphery of the cell. There may be regions of dividing cells that are so densely packed with cytoskeletal elements that large organelles are effectively excluded from them. The most extensive data I know of are presented in Alberts *et al.* (1983, Table 7-1), for liver hepatocytes. The mitochondria occupy 22% of the cell volume, but 23% is occupied by the nucleus, endoplasmic reticulum, Golgi apparatus, lysosomes, and peroxisomes. There are no data for the cytoskeleton in these same cells, but in cultured animal cells, the microfilaments, intermediate filaments, microtubules, and microtrabeculae occupy 16−21% (Gershon *et al.*, 1985). After 22% and 21% are subtracted for other organelles, the mitochondria may still occupy less than 40% of the available volume in animal cells. Unfortunately, no firm conclusions can be drawn from these data, because there are no numerical data on mitochondrial partitioning in animal cells other than scorpion spermatocytes.

Second, our mathematical treatment was necessarily over-simplified because it considered only the volume of the organelles, not their shape. *Figure 5* uses two-dimensional cells and organelles to show that shape is important. In a square cell, four organelles can all fit in one half of the cell, and potentially be partitioned into one daughter cell, if they are rectangular. But four round organelles of the same volume cannot; in fact in this case, the four round organelles can only be partitioned equally (2:2) because not even three can fit in one half of the cell. Although the figure is two-dimensional, the same principle obviously applies to real, three-dimensional cells. The equations of Birky and Skavaril (1984) are strictly applicable only to perfectly elastic organelles or cells. The correct mathematics will be different for each organism because

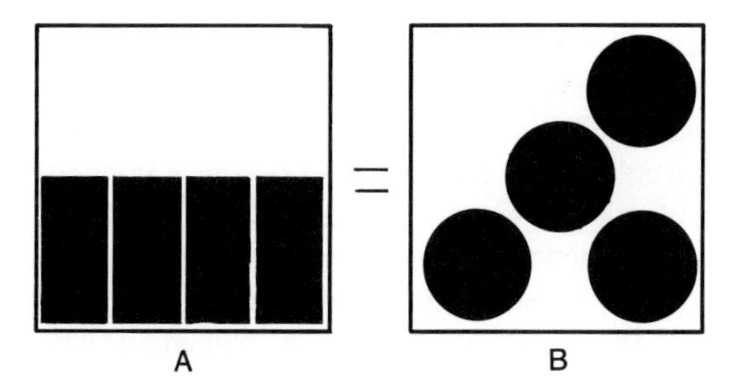

Figure 5. A two-dimensional representation of the joint effect of shape and volume on the partitioning of organelles. In the square cell on the left, four square organelles can all fit in one half of the cell and hence could all be included in the same daughter cell after cytokinesis. But not even three round organelles of the same volume can fit in half of the cell; partitioning in this case is always equal.

the shapes of the cells and organelles are so variable. A more realistic and detailed mathematical treatment can only be done for specific cases and only with considerable difficulty or oversimplification.

This is probably unnecessary, because the available theory can be used to show that volume effects alone cannot account for the partitioning data. Taking the observed frequencies of equal partitioning from the studies of chloroplast partitioning in *Olisthodiscus* and clover, and of mitochondria in scorpions, the equations of Birky and Skavaril (1984) have been used to determine the corresponding ratios of organelle volume to cell volume. In the latter two cases, the observed results were impossible, according to the equations. The frequency of equal partitioning was so large that the equations predicted that unequal partitioning would be impossible, and yet it occurred. Using *Figure 4* as an example, the data lie directly on the vertical part of the curve for the equal partitioning case. Only in the case of *Olisthodiscus* do the observed data lie within the parameter space of the equations, that is, within the range of organelle numbers and volume ratios where unequal partitioning as well as equal partitioning is possible. Obviously this does not mean that the data are wrong, or that the equations are incorrect. Nor does it imply that volume effects play no role in partitioning. It does mean that it is impossible to explain the results by volume effects alone, even if we knew the available volume as discussed above. Clearly the volumes of the organelles are not so great as to prevent more than half of them from entering the same cell, but the equations say that they behave as if they were. Moreover, if equations were available that took the true shape of the cells and organelles into account, as discussed above, this would only make matters worse by further reducing the parameter space of the equations. In fact, it is likely that such equations would show that volume effects are not adequate to explain the *Olisthodiscus* data, either, since they barely fit within the parameter space of the existing equations.

These results have even wider implications, because they imply that any conceivable partitioning mechanism that is infallible cannot provide a complete explanation of the data. All such models can probably be treated with the equations of Birky and Skavaril (1984) by defining an *effective* volume ratio, which is the ratio of organelle volume to cell volume that would explain the observed distributions. We have seen that no volume, real or effective, is sufficient to explain the data. For example, it has been suggested that organelles repel each other, moving so as to position themselves equidistantly. This would amount to assigning each organelle an effective volume, larger than the volume that it actually occupies, like the 'personal space' that human beings try to protect by behavioural patterns. But the argument presented here says that the effective volume varies randomly in size and, like our own personal space, can be violated under some circumstances.

Localization of organelles by the cytoskeleton

Since volume effects and variants thereof cannot explain the partitioning of multi-copy organelles, we are forced to look to the cytoskeleton. It is now well known that organelles can be moved along elements of the cytoskeleton, and there is reason to believe that most or all organelle movements happen this way (e.g., Koonce *et al.*, 1986). A good review of the background of this subject can be found in Alberts *et al.* (1983). It is reasonable to hypothesize that organelles are moved by the cytoskeleton to positions where they will tend to be partitioned non-randomly, and that they may also be anchored in those positions by the cytoskeleton. This hypothesis has not been tested, but even if it proves to be true, it still would not explain why the organelles are moved, and anchored, to positions where they will tend to be partitioned equally as opposed to unequally. At the present state of our knowledge, the cytoskeleton is like an automobile whose driver cannot be interviewed; we can use the engine to explain how it moves and the parking brake to explain why it stays there, but we have no way of explaining why it goes some places and not others. A great deal of progress is being made in the analysis of the cytoskeleton, but most of it is being done with animal cells rather than microorganisms. Microbiologists need an experimental system, a cell which can be induced to undergo mitosis at will, which contains a modest number of organelles whose movements can be followed, and which can be experimentally manipulated. Useful kinds of manipulations might include disruption of specific components of the cytoskeleton by drugs, mechanical rearrangements of the mitotic spindle, and microinjection of organelles.

Summary

There have been enough qualitative studies of the partitioning of mitochondria and chloroplasts at cell division to tell us that many different patterns can be found, even within major taxa. Consequently we are unlikely to be able to identify a general mechanism for partitioning organelles non-randomly. There have been very few quantitative studies of organelle partitioning, and many more are needed. The available data suggest that cells do not partition their organelles with the same degree of regularity that chromosomes are partitioned, but neither are the cytoplasmic organelles partitioned strictly at random. Some of the non-randomness can probably be explained simply by taking into account the volumes occupied by the organelles, but this is almost certainly not a complete explanation. It is likely that the cytoskeleton positions organelles so that they are partitioned more uniformly, but this has not been verified.

References

Alberts,B., Bray,D., Lewis,J., Raff,M., Roberts,K. and Watson,J.D. (1983) *Molecular Biology of the Cell.* Garland Publishing, New York.

Anton-Lamprecht,L. (1967) Anzahl und Vermehrung der Zellorganellen im Scheitelmeristem von *Epilobium. Berichte deutsche botanische Gesellschaft,* **80**, 747–754.

Aufderheide,K. (1979) Mitochondrial associations with specific microtubular components of the cortex of *Tetrahymena thermophila.* I. Cortical patterning of mitochondria. *Journal of Cell Science,* **39**, 299–312.

Berger,J.B. (1979) Regulation of macronuclear content in *Paramecium tetraurelia. Journal of Protozoology,* **26**, 18–28.

Birky,C.W.,Jr (1983) The partitioning of cytoplasmic organelles at cell division. *International Review of Cytology, Supplement,* **15**, 49–89.

Birky,C.W.,Jr and Skavaril,R.V. (1984) Random partitioning of cytoplasmic organelles at cell division: the effect of organelle and cell volume. *Journal of Theoretical Biology,* **106**, 441–447.

Butterfass,T. (1969) Die Pastidenverteilung vei der Mitose der Schleisszellenmutterzellen von haploidem Schwedenklee (*Trifolium hybridum* L.). *Planta,* **84**, 230–234.

Doerder,F.P. (1979) Regulation of macronuclear DNA content in *Tetrahymena thermophila. Journal of Protozoology,* **26**, 28–35.

Gershon,N.D., Porter,K.R. and Trus,B.L. (1985) The cytoplasmic matrix: its volume and surface area and the diffusion of molecules through it. *Proceedings of the National Academy of Science USA,* **82**, 5030–5034.

Hall,W.T. and Claus,G. (1963) Ultrastructural studies on the blue-green algal symbiont in *Cyanophora paradoxa* Korschikoff. *Journal of Cell Biology,* **19**, 551–563.

Hennis,A.A. and Birky,C.W.,Jr (1984) Stochastic partitioning of chloroplasts at cell division in the alga *Olisthodiscus*, and compensating control of chloroplast replication. *Journal of Cell Science,* **70**, 1–15.

Hood,R.D., Watson,O.F., Deason,T.R. and Benton,C.L.B.,Jr (1972) Ultrastructure of scorpion spermatozoa. *Cytobios,* **5**, 167–177.

Hussein,Y., Heizmann,P., Nicholas,P. and Nigon,V. (1982) Quantitative estimations of chloroplast DNA in bleached mutants of *Euglena gracilis. Current Genetics,* **6**, 111–117.

Kuroiwa,T., Kawano,S. and Hizume,M. (1977) Studies on mitochondrial structure and function in *Physarum polycephalum.* V. Behavior of mitochondrial nucleoids throughout mitochondrial division cycle. *Journal of Cell Biology,* **72**, 687–694.

Koonce,M.P., Euteneuer,U. and Schliwa,M. (1986) *Reticulomyxa*: a new model system of intracellular transport. *Journal of Cell Science, Supplement,* **5**, 145–159.

Stevens,B.J. (1977) Variation in number and volume of the mitochondria in yeast according to growth conditions. A study based on serial sectioning and computer graphics reconstitution. *Biologie Cellulaire,* **28**, 37–56.

Trench,R.K., Pool,R.R.,Jr, Logan,M. and Engelland,A. (1978) Aspects of the relation between *Cyanophora paradoxa* (Korschikoff) and its endosymbiotic cyanelles *Cyanocyta korschikoffiana* (Hall and Claus). I. Growth, ultrastructure, photosynthesis and the obligate nature of the association. *Proceedings of the Royal Society of London B,* **202**, 423–443.

Wasmann,C.C., Loeffelhardt,W. and Bohnert,H.J. (1987) Cyanelles: organization and molecular biology. In *Cyanobacteria.* Fay,P. and Van Baalen,C. (eds), Elsevier Science Publishers, Amsterdam, 303–324.

Whatley,J.M. (1976) Bacteria and nuclei in *Pelomyxa palustris*: comments on the theory of serial endosymbiosis. *New Phytologist,* **76**, 111–120.

Whatley,J.M. (1983) Plastids—past, present, and future. *International Review of Cytology, Supplement,* **14**, 329–373.

Wilson,E.B. (1925) *The Cell in Development and Heredity.* 3rd edition. MacMillan, New York.

Wilson,E.B. (1931) The distributon of sperm forming materials in scorpions. *Journal of Morphology and Physiology,* **52**, 429–483.

CHAPTER 5

Cell polarity and differentiation in cellular slime moulds

D.J.WATTS

Biochemistry Department, University of Sheffield, Sheffield S10 2TN, UK

Introduction

There are two phases – growth and development – in the life cycle of the cellular slime moulds. During development, which is induced by starvation, the slime mould amoebae collect into aggregates usually in response to a chemoattractant synthesized and secreted by the amoebae. The aggregates then pass through a series of morphogenetic stages until finally transforming into fruiting bodies composed of stalk cells and spores. Development is therefore accompanied by a period of cell differentiation during which each amoeba has to differentiate into either a stalk cell or a spore. Because differentiation in cellular slime moulds involves formation of only two specialized cell types, it is much simpler than cell differentiation in higher organisms where many different cell types are formed. In consequence, the cellular slime moulds are much studied as 'model systems' of development and differentiation.

Dictyostelium discoideum is the species of cellular slime mould that has been most studied and this chapter will therefore be mainly concerned with its development and differentiation. In the laboratory, it is possible to maintain amoebae of *D. discoideum* in conditions that ensure that all the amoebae pass synchronously through the various stages of development and reach any particular developmental stage at a predictable time (*Figure 1*) after development has been initiated by depriving the amoebae of nutrients (Sussman and Lovgren, 1965). In these conditions, fruiting bodies appear after about 24 h development although the spores are probably not mature until after 30 h.

At about 11 h development, each hemispherical aggregate becomes a polar structure as it forms an apical tip which will then persist throughout the remaining stages of development. However, it is particularly at the migrating slug stage of development (*Figure 2*) that a relationship between aggregate polarity and cell differentiation has been clearly established. The migrating slug has an obvious anterior–posterior polarity and moves towards higher temperature and light intensity. It has been shown that the cells that will eventually form the stalk cells of the fruiting body make up the anterior 20–25% of the slug whereas the cells in the remaining, posterior part of the slug will eventually differentiate into spores (Raper, 1940). Furthermore, although the cells in the migrating slug are still all amoeboid, cell differentiation has already begun since the anterior, prestalk cells differ from the posterior, prespore cells in enzyme com-

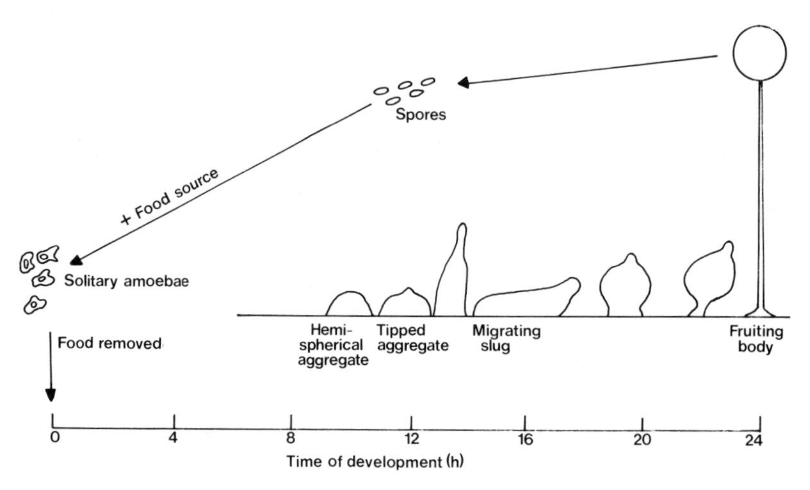

Figure 1. Life cycle of *Dictyostelium discoideum*. The time course for the developmental stages applies to the laboratory conditions described by Sussman and Lovgren (1965).

position (Newell *et al.*, 1969; Tsang and Bradbury, 1981), mRNA composition (Mehdy *et al.*, 1983; Barklis and Lodish, 1983; Datta *et al.*, 1986), pattern of protein synthesis (Alton and Brenner, 1979; Ratner and Borth, 1983; Morrissey *et al.*, 1984; Kopachik *et al.*, 1985), ability to incorporate [^{3}H]fucose into glycoproteins (Gregg and Karp, 1978), staining properties with vital stains (Bonner, 1952) and ultrastructure (Hohl and Hamamoto, 1969; George *et al.*, 1972). The differences in ultrastructure have been particularly helpful in distinguishing the prespore cells from the prestalk cells. The former cells contain unique structures called prespore vacuoles (Hohl and Hamamoto, 1969; Gregg and Badman, 1973) in which the spore coat proteins [and possibly a mucopolysaccharide specific to prespore cells; Takeuchi (1972)] are stored (Devine *et al.*, 1983) before eventually being released to the exterior of the cell where they help to form the resistant spore coat. Antibodies raised against spores of *Dictyostelium mucoroides* will bind to the prespore vacuoles (Takeuchi, 1972; Forman and Garrod, 1977a) and, if fluorescent antibodies are used to stain sections of migrating slugs, the posterior prespore cells become highly fluorescent whereas the prestalk cells at the anterior show no more than a slight autofluorescence (*Figure* 3). Prespore cell differentiation is therefore often identified by making use of the spore antiserum.

'Positional information'

The apparent relationship in *D. discoideum* between cell fate and the anterior–posterior polarity of a migrating slug has made it seem probable that cell differentiation is controlled by cell position within the aggregate. This would require that cells should remain undifferentiated until a polar aggregate has formed and that only then can cell differentiation begin with cells at the anterior of the aggregate being induced to follow the stalk pathway of differentiation and cells at the posterior being induced to initiate spore differentiation. This description of slime mould differentiation and pattern formation generally also requires that a concentration gradient of a chemical (morphogen)

2

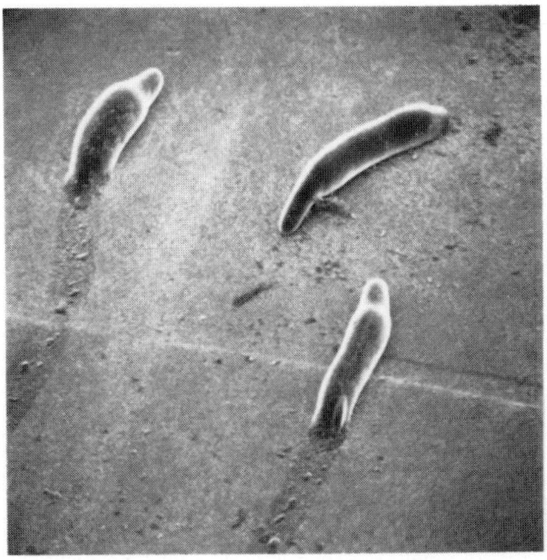

3

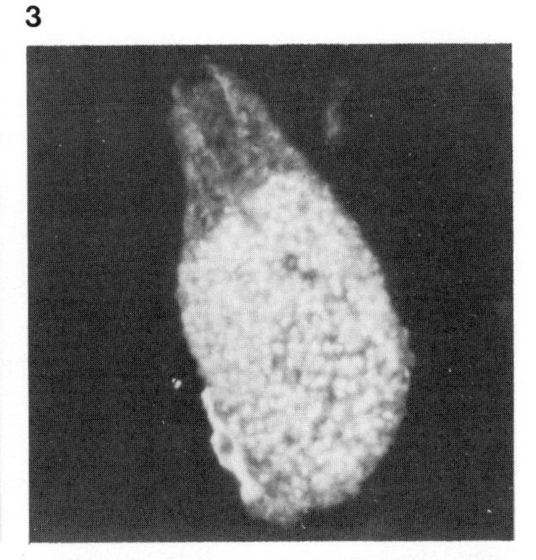

Figure 2. Migrating slugs of *Dictyostelium discoideum*. Amoebae were allowed to develop on cellulose acetate filters. When migrating slugs had formed, the filters were frozen in melting Freon 22 and were examined under a scanning electron microscope without further treatment. The slugs had migrated randomly because they had been maintained at a constant temperature (22°C) in the dark. The anterior tips are prominent. (Micrograph by T.E.Treffry and D.J.Watts.) Further scanning electron micrographs of *D. discoideum* are shown in Treffry and Watts (1976) and Watts and Treffry (1976). Approximate magnification ×30.

Figure 3. The prestalk–prespore pattern in a slug. The section of a slug was stained with a spore antiserum (Forman and Garrod, 1977a) so that the prespore vacuoles in the prespore cells became highly fluorescent. (Micrograph provided by D.R.Garrod.) Approximate magnification ×200.

should be set up along the anterior–posterior axis of the slug. The morphogen could then provide the undifferentiated cells with 'positional information' (Wolpert, 1969) since any position along the aggregate would be characterized by a particular concentration of morphogen. Thus a cell could determine its position by monitoring the local concentration of morphogen and respond by following the appropriate pathway of differentiation.

The hypothesis is illustrated in its simplest form in *Figure 4* in which it is assumed that a stalk-inducing morphogen is synthesized at the anterior of the slug and, having diffused along the slug, is destroyed at the posterior so that a concentration gradient is created. Cells at the anterior of the slug, where the concentration of morphogen is high, will therefore be induced to differentiate into stalk cells. At the posterior, where the morphogen concentration is too low to induce stalk differentiation, the cells can eventually differentiate into spores. A threshold concentration of morphogen will therefore specify the position of the boundary between the prestalk and prespore cells; indeed, transmission electron micrographs (Schaap *et al.*, 1982) have confirmed that, as indicated in *Figure 3*, the boundary is extremely sharp in *D. discoideum* slugs. As shown in *Figure 4*, it would also be expected, on the basis of this description of the model, that the proportions of prestalk and prespore cells in a slug would be independent of the size of the slug; this is approximately what has been found experimentally (Williams *et al.*, 1981).

One difficulty with models based on positional information is to provide a plausible explanation for the generation of the morphogen concentration gradients. However,

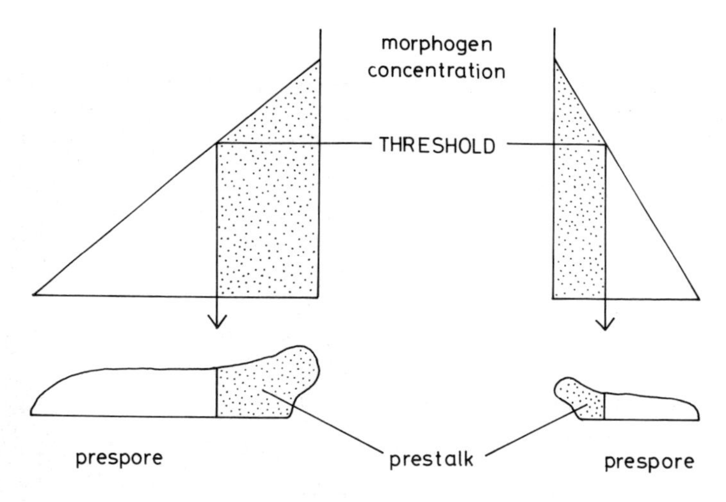

Figure 4. Positional information in *Dictyostelium discoideum*. It is assumed that a stalk-inducing morphogen is synthesized at the anterior of the slug and destroyed at the posterior so that a concentration gradient of morphogen is created. The gradient will therefore be steeper in a small slug than in a larger slug and, as a result, the threshold concentration of morphogen sets the prestalk–prespore boundary so that, in each slug, there are approximately similar proportions of prestalk and prespore cells.

MacWilliams and Bonner (1979) and Gross *et al.* (1981) have proposed that one form of the model described by Gierer and Meinhardt (1972) may account for gradient formation in *D. discoideum* although, more recently, Meinhardt (1983) has expressed reservations about this. According to MacWilliams and Bonner (1979), the appropriate form of the Gierer and Meinhardt (1972) model predicts that two morphogens – an activator and an inhibitor – would be required to account for pattern formation in *D. discoideum*. Synthesis of the activator has to be autocatalytic so that any tendency for a cell to synthesize activator, perhaps in response to a local perturbation in its environment, will lead rapidly to greatly increased synthesis of activator by that cell and also by neighbouring cells as activator diffuses into them. Synthesis of activator has also to stimulate synthesis of the inhibitor which, in turn, antagonizes the autocatalysis of activator synthesis and so moderates synthesis of the activator. If the rates of activator and inhibitor synthesis and degradation are such that the inhibitor is able to diffuse further along the aggregate than the activator, a concentration gradient of activator will be created. Furthermore, this model offers the possibility that there could be the sudden, precipitous fall in activator concentration that is presumably needed to define the sharp prestalk–prespore boundary found in a migrating slug. The inhibitor would also be distributed in a concentration gradient because its synthesis is sensitive to the concentration gradient of activator, but the gradient in inhibitor concentration would be less steep than that in activator concentration. Finally, in this model, formation of the concentration gradient of activator would allow not only cell fate, and thus the prestalk–prespore pattern, to be controlled by positional information but would, in itself, be the mechanism creating aggregate polarity. Thus the original site of activator synthesis would become one end of the polar aggregate.

D. discoideum morphogens

To confirm that cell differentiation in *D. discoideum* is controlled by positional information, it would appear essential to show, at the very least, that the amoebae synthesize stalk- or spore-inducing morphogens and that these morphogens are distributed in suitable concentration gradients within the polar aggregates. Recently, a number of compounds have been identified which, because of their effects on amoebal differentiation, appear to act as *D. discoideum* morphogens and this has greatly added to the attraction of studying differentiation in *D. discoideum* since the morphogens controlling cell differentiation in most other systems remain elusive.

The morphogens that have been fully characterized are 3′,5′-cyclic AMP, Differentiation Inducing Factor (DIF), ammonia and adenosine. In addition, a spore-inducing morphogen has been isolated but its chemical structure is unknown (Wilkinson *et al.*, 1985; Kumagai and Okamoto, 1986).

Cyclic AMP. This has been identified as the chemoattractant controlling aggregation of *D. discoideum* amoebae (Konijn *et al.*, 1967). However, cyclic AMP signalling systems are induced, following cell aggregation, in species of cellular slime moulds that do not use cyclic AMP as a chemoattractant during aggregation (Schaap and Wang, 1984), suggesting that cyclic AMP is also involved in controlling post-aggregation events in the cellular slime moulds. A possible role for cyclic AMP in cell differentitation was first described by Bonner (1970) who found that, when *D. discoideum* amoebae were at low population density on the substratum, and were therefore unable to aggregate, some of the amoebae would differentiate into stalk cells in the presence of high cyclic AMP concentrations. This suggested that cyclic AMP might be a stalk-inducing morphogen and this appeared to be confirmed by later studies described by Town *et al.* (1976). They found that the strain V12M2 of *D. discoideum* is particularly responsive to cyclic AMP and, when incubated in tissue culture dishes in a suitable buffer solution containing 1 mM cyclic AMP, all the amoebae of this strain eventually differentiated into stalk cells in the absence of morphogenesis. Wild-type cells of strain V12M2 never differentiated *in vitro* into spores but mutant, sporogenous strains have been isolated that will give rise to both spores and stalk cells *in vitro* (Town *et al.*, 1976). A high cyclic AMP concentration is still then required to induce spore differentiation (Town *et al.*, 1976; Kay, 1982) and this has cast doubt on the proposal that cyclic AMP is exclusively a stalk-inducing morphogen. Furthermore, recent studies have indicated that the cyclic AMP relay in slugs may be primarily concerned with control of prespore cell differentiation (Schaap and Wang, 1986). Perhaps also consistent with this are studies of differentiation *in vitro* of wild-type amoebae of strain V12M2 because it has been found that, before giving rise to stalk cells, all the amoebae first differentiate, under the influence of cyclic AMP, into prespore cells (Kay *et al.*, 1978).

Differentiation inducing factor. It has become clear that cyclic AMP induces stalk cell differentiation *in vitro* because it stimulates the amoebae, at 9 – 10 h development, to initiate synthesis and secretion of DIF which appears to be a stalk-inducing morphogen (Town *et al.*, 1976; Town and Stanford, 1979; Brookman *et al.*, 1982; Sobsolewski *et al.*, 1983; Kopachik *et al.*, 1983; Kay and Jermyn, 1983). A mutant strain has been isolated (Kopachik *et al.*, 1983) that stops development when hemispherical aggregates have formed and fails to accumulate DIF. However, the mutant amoebae remain responsive to DIF and, in its presence, will differentiate into stalk cells. Purification of DIF has shown that it comprises a small family of compounds of low molecular weight (Kay *et al.*, 1983); although their chemical structures are known, details have yet to be published.

Ammonia. During development, amoebae resort to amino acid degradation as the major source of energy production and large amounts of ammonia are therefore excreted. Ammonia can affect development; a high concentration of ammonia in the environment induces a long period of slug migration and thus delays the subsequent stages of development and differentiation (Schindler and Sussman, 1977). This has led to the proposal (Sussman *et al.*, 1977; Sussman and Schindler, 1978) that ammonia may have effects on cell differentiation and that these might occur because ammonia inhibits cyclic AMP accumulation in amoebae (Schindler and Sussman, 1979). It was also proposed that the slime sheath surrounding the migrating slug might be more permeable to ammonia at the anterior than at the posterior of the slug and that this could result in ammonia becoming distributed in a concentration gradient along the slug and, through its effect on cyclic AMP accumulation by amoebae, induce formation of a concentration gradient of cyclic AMP that could control cell differentiation and pattern formation. Although more recent studies make it seem improbable that ammonia could have effects on cell differentiation by controlling cyclic AMP concentration in the slug, they have confirmed that ammonia does affect cell differentiation (Kay, 1979; Gross *et al.*, 1981; Neave *et al.*, 1983; Dominov and Town, 1986). Thus, spore differentiation *in vitro* (where there is a uniformly high cyclic AMP concentration) by sporogenous mutants is favoured by high pH and high ammonia concentration, whereas low pH and low ammonia concentration favour stalk cell differentiation. The apparently opposed effects of DIF and ammonia on cell differentiation have suggested (Gross *et al.*, 1981) that DIF may be the activator, and ammonia the inhibitor, in the activator-inhibitor model for pattern formation that has been described earlier. This would imply that ammonia should be at highest concentration at the anterior of the slug whereas the theory proposed by Sussman *et al.* (1977) and Sussman and Schindler (1978) predicts that ammonia concentration should be lowest in this region.

Adenosine. Although it has been known for some years that formycin B, which is an analogue of adenosine, affects the post-slug stages of development (Brackenbury *et al.*, 1974) and that amoeboe have cell surface receptors for adenosine (Newell, 1982; Newell and Ross, 1982), it was only recently that a possible role for adenosine as a morphogen was identified (Schaap and Wang, 1986). When migrating slugs were maintained on agar containing 5−10 mM adenosine, the normal prestalk−prespore pattern was disrupted and a thickish layer of prestalk cells was formed at the undersurface of the slug below the prespore cells whilst the anterior, prestalk region was reduced in size. Further evidence that adenosine is probably involved in stalk differentiation *in vivo* was obtained when it was found that the anterior cells started to synthesize prespore antigens (detected by using spore antiserum), and thus to change their differentiated state, when slugs were incubated in adenosine deaminase which would destroy endogenous adenosine. Overall, Schaap and Wang (1986) concluded that adenosine is an inhibitor of prespore cell differentiation and probably acts by preventing cyclic AMP from inducing prespore cell differentiation.

Concentration gradients of morphogens. There are considerable difficulties in determining whether morphogens are distributed in concentration gradients in migrating slugs because the slugs, at best, are only a few millimetres long and contain only tiny amounts of morphogens. Nevertheless, the distributions of cyclic AMP, DIF and ammonia have been determined.

Several groups (Garrod and Malkinson, 1973; Pan *et al.*, 1974; Brenner, 1977) have investigated the cyclic AMP concentration in slugs and all agree that the concentration is higher at the anterior than at the posterior. However, quantitative measurements (Garrod and Malkinson, 1973; Brenner, 1977) have indicated that the cyclic AMP concentration in the prestalk region of the slug is no more than twice the concentration in the prespore region. Any gradient in cyclic AMP concentration must, therefore, be extremely shallow and, as pointed out by Brenner (1977), it is difficult to conceive how such a gradient could control cell differentiation so as to create the sharp prestalk–prespore boundary existing in a migrating slug. The concentration of DIF in the prestalk region of the slug has also been found to differ from that in the prespore region by about a factor of only two but, paradoxically, the higher concentration of the stalk-inducing morphogen is in the prespore region of the slug (J.J.Brookman, personal communication). This may indicate that DIF is only the precursor of the stalk-inducing morphogen (Meinhardt, 1983); conversion of DIF into the active morphogen at the anterior, but not at the posterior, of the slug would then lead to depletion of DIF at the anterior of the slug and thus to the concentration gradient of DIF found experimentally. Furthermore, the active, stalk-inducing morphogen would then be at highest concentration in the prestalk region of the slug. However, this argument implies that the morphogen possibly derived from DIF cannot be involved in control of the initial steps of cell differentiation since its distribution would itself be dependent on prior cell differentiation (see the discussion below on adenosine). There is no evidence for a concentration gradient of ammonia in migrating slugs (Wilson and Rutherford, 1978) although, at later stages of development, large differences have been detected between the ammonia concentrations in the spore- and stalk-forming regions of aggregates.

The adenosine concentration in slugs has yet to be determined. However, it has been assumed that the precursor of the adenosine is extracellular cyclic AMP and it is known that the two enzymes (3′,5′-cyclic AMP phosphodiesterase and 5′-nucleotidase) needed to catalyse its conversion into adenosine have the required properties of being bound to the plasma membranes of the amoebae and of being able to utilize extracellular substrates (Rossomando and Cutler, 1975; Armant *et al.*, 1980; Schaap and Spek, 1984). The phosphodiesterase is absent from amoebae during growth but is synthesized during early development. Its activity is controlled by an inhibitor protein that is also synthesized during development (Gerisch *et al.*, 1972) and it would appear that the inhibitor is distributed in the migrating slug so that the phosphodiesterase has highest activity in the prestalk region (Brown and Rutherford, 1980; Tsang and Bradbury, 1981; Schaap and Spek, 1984; Mee *et al.*, 1986). Activity of the 5′-nucleotidase can be detected even in growing amoebae but is also controlled by a tightly-bound inhibitor (Das and Weeks, 1980, 1981). In the migrating slug, this inhibitor is distributed such that the enzyme has its highest activity in the prestalk region (Armant and Rutherford, 1979, 1981). It would, therefore, seem highly probable that adenosine will be at highest concentration in the prestalk region of a slug because the concentration of the cyclic AMP precursor and the activities of the two enzymes catalysing its conversion into adenosine are all highest in this region. Nevertheless, it must be recognized that these arguments indicate that adenosine cannot be of primary importance in controlling prestalk cell differentiation; presumably it is the factors bringing about the differential distributions of the phosphodiesterase and 5′-nucleotidase inhibitors in the amoebae of the polar aggregates that have this role.

Clearly, cyclic AMP, DIF, ammonia and adenosine have marked effects on differentiation of *D. discoideum* amoebae but studies of the distributions of these morphogens in slugs appear to cast doubt on any hypothesis proposing that such morphogens control the initial stages of differentiation when amoebae start to follow the pathway leading to spore formation or the pathway leading to stalk cell formation. It remains possible, of course, that differentiation is initially controlled by concentration gradients of morphogens that have still to be identified or by concentration gradients of the known morphogens that are not detected when the overall (that is, intracellular plus extracellular) concentrations of these morphogens are measured. The latter argument may particularly apply to DIF (Neave *et al.*, 1986). Nevertheless, studies of the distributions of the known morphogens in slugs would suggest that the hypothesis – that there is a relationship between cell fate and cell position in an aggregate because cell differentiation is controlled by positional information – should be questioned and that there could be an alternative explanation for the relationship.

Slug polarity and cell sorting

A relationship between cell differentiation and aggregate polarity would arise if amoebae initiate differentiation early in development and then sort out so that a polar aggregate is created as cells of the same type (presumptive spore cells or presumptive stalk cells) collect together. Aggregate polarity would then be a consequence – not a cause – of cell differentiation. A number of different experimental approaches have provided support for this idea.

(i) If a migrating slug is dissociated into a random mixture of prespore and prestalk cells, it is possible for the cells to sort out and to give rise to a new slug in which the prestalk cells of the original slug return to the anterior of the new slug and the prespore cells of the original slug return to the posterior of the new slug (Takeuchi, 1969; Sternfeld and David, 1981). It is therefore apparent that cells that have begun to differentiate can, by sorting out, create a polar aggregate.

(ii) Studies of cell differentiation *in vitro* (Town *et al.*, 1976) have indicated that spore and stalk cell differentiaion can occur in the absence of morphogenesis. There is therefore no requirement for amoebae to be in a polar aggregate in order to initiate differentiation.

(iii) When amoebae are suspended in phosphate buffer and shaken gently, they rapidly form approximately spherical aggregates, each of which becomes surrounded by a slime sheath. By using a spore antiserum for immunofluorescent staining of sections of aggregates taken at various times of development it is possible to show that many of the amoebae in the aggregates differentiate into prespore cells. It is assumed that the unstained amoebae are the precursors of stalk cells. At first, the two cell types are randomly distributed within the aggregates but eventually they sort out. The stalk precursor cells then collect together in a small region at the surface of the aggregate and the rest of the aggregate comprises only prespore cells (Forman and Garrod, 1977b; Takeuchi *et al.*, 1982). The resulting pattern of the two cell types in these aggregates, which have no anterior tip or other sign of anterior–posterior polarity, is very similar to the prestalk–prespore pattern in migrating slugs and implies that pattern formation in a migrating slug may also result from sorting out of already differentiating cells and may not be dependent on the polarity of the aggregate.

Early stages of cell differentiation

To explore further the implications of the experiments described above, which were mainly carried out with amoebae in abnormal conditions of development in which they were unable to complete formation of fruiting bodies, it seemed necessary to investigate the early stages of differentiation of amoebae maintained in more natural conditions of development. However, such studies were hampered by a lack of criteria by which the initial stages of differentiation could be identified. For example, the spore antiserum could be used only after 10 h development to detect prespore cell differentiation (Forman and Garrod, 1977a). Nevertheless, it appeared probable that, if differentiation gives rise to two cell types very early in development, the cells would have to differ in surface properties in order to be able to sort out, after aggregates had formed, to give the prestalk–prespore pattern. It therefore seemed worthwhile to investigate the early stages of development and differentiation by making use of a technique (thin-layer counter-current distribution in an aqueous, two-phase, polymer system – TLCCD) that will separate cells having different cell surface properties and thus different plasma membrane structures.

Counter-current distribution of cell types

The TLCCD technique (see Sharpe, 1984) is based on the tendency of mixtures of aqueous solutions of Dextran T500 and poly(ethyleneglycol) 4000 (or 6000) to separate into two phases on standing. Cells with different surface properties will have different affinities for the two phases and, when added to an appropriate two-phase system, will tend to separate. If a phase system is used in which there is no potential difference between the upper and lower phases, cells will partition between the two phases owing to differences in cell surface hydrophobicity; the cells with the more hydrophobic cell surface properties will partition into the upper, poly(ethyleneglycol)-rich phase whereas the cells with less hydrophobic cell surface properties will partition into the lower, Dextran-rich phase or, more usually, collect at the interface. In order to obtain useful

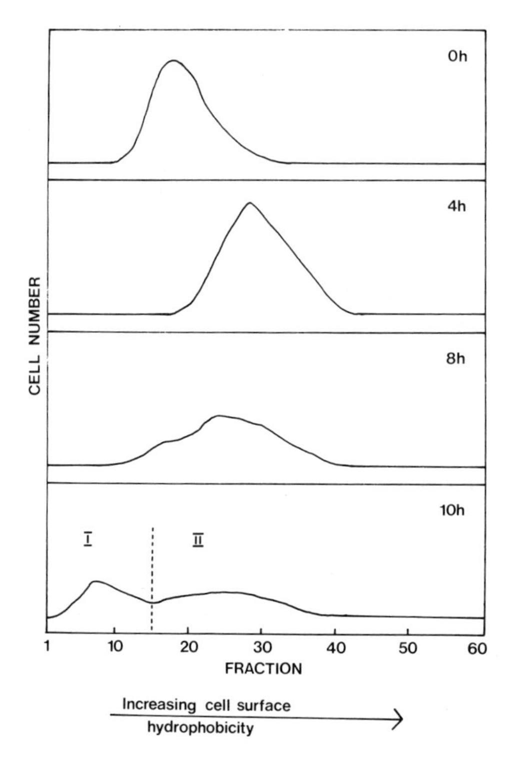

Figure 5. TLCCD profiles of developing *Dictyostelium discoideum* amoebae. Amoebae were taken at the indicated times of development for TLCCD. The counter-current distributions involved the amoebae being partitioned sixty times in a Dextran–poly(ethyleneglycol), two-phase system. At 10 h development, two populations of amoebae (peaks I and II) were present.

cell separation, it is usually necessary to repeat the partitioning many times (for example, sixty times) with fresh upper and lower phases in a countercurrent distribution apparatus (see Sharpe, 1984). Like many other cell types, amoebae of *D. discoideum* remain fully viable during partitioning.

Amoebae harvested during the first seven hours of development partitioned as a single, fairly sharp peak (*Figure 5*) (Sharpe *et al.*, 1982) and this showed that, at any particular stage of early development, all the amoebae had similar, though not identical (Sharpe and Watts, 1984), cell surface properties. However, the position of the peak changed with time of development and it was apparent that, during the first four hours of development, the surface properties of the amoebae became more hydrophobic and then, at 6 h development, more hydrophilic. The changes in the surface properties of the amoebae appeared consistent with the report (Toda *et al.*, 1980) that there are marked changes in the glycoprotein composition of the amoebal plasma membranes during early development. Between 8 h and 11 h development, the surface properties of the amoebae became extremely heterogeneous (*Figure 5*) as a proportion of the amoebae gained more hydrophilic cell surface properties, and, at 9–11 h development (11 h is the latest stage that can be investigated by TLCCD), it was clear that two different populations of amoebae had formed (designated I and II in *Figure 5*). When amoebae from peak I were mixed with an equal number of amoebae from peak II and allowed to form fruiting bodies, it was found that all the amoebae from peak I formed spores whereas the stalk cells were derived from the amoebae in peak II (Sharpe *et al.*, 1982; Sharpe *et al.*, 1984). Furthermore, when amoebae from peak I were allowed to complete development alone, they behaved like prespore amoebae isolated from migrating slugs since they formed fruiting bodies having more or less normal proportions of spore and stalk cells. By contrast, amoebae from peak II formed fruiting bodies with abnormally long stalks and this behaviour was similar to that of prestalk cells isolated from slugs. It was concluded that peak I comprised presumptive spore cells and peak II presumptive stalk cells. At 10 h development, about 40% of the amoebae were in peak I (presumptive spore cells) whereas at least 70% of the cells in a fruiting body are spores. Some of the amoebae distributing in peak II must therefore also eventually differentiate into spores.

These results have been criticized on the basis that TLCCD takes 10 h for completion so that, even though the amoebae are maintained at 5°C, the changes in amoebal surface properties may be created artificially during the long period of cell separation. This seemed improbable since the TLCCD results were extremely reproducible and, more recently, it has been confirmed that the two populations of amoebae with markedly different surface properties must be formed during development, and not during TLCCD, because it has been possible to isolate amoebae from these two populations at 10–11 h development by using a modified partitioning technique in which the amoebae remain in the phase systems at 5°C for no more than 30 min (D.J.Watts, G.M.Knight and P.T.Sharpe, unpublished results).

The polypeptide compositions of the amoebae from peak I or peak II at 10 h development were investigated by two-dimensional, polyacrylamide-gel electrophoresis (Sharpe *et al.*, 1985). Eight polypeptides were detected that had been gained during development uniquely by amoebae in peak I and there were five polypeptides that had been lost uniquely by these amoebae. Similarly, there were gains (four) and losses (three) of polypeptides that were specific to the amoebae in peak II. These results, together with those indicating that cell-specific changes in surface properties also occur during the first 10 h of development, confirmed the conclusion, tentatively reached previously by Forman and Garrod (1977a), that cell differentiation is initiated no later than 10 h development. Since, at this time of development, aggregates have no apical tips, or other signs of polarity, it seemed reasonable to conclude that, even in normal conditions of development, initiation of cell differentiation is not dependent on aggregate polarity. This conclusion was further supported by an electronmicroscopic study of aggregates at 10 h development (Schaap *et al.*, 1982). On examination of sections of the aggregates stained with OsO_4, electron-dense and electron-lucent cells were found to be present and it appeared that the former were the presumptive spore cells. The two cell types were randomly distributed within a 10 h aggregate and this confirmed that initiation of cell differentiation cannot be in response to positional information. Similarly, Tasaka *et al.* (1983) have used monoclonal antibodies to identify the presumptive spore and stalk cells in sections of early aggregates and have also found that the two cell types are randomly distributed within the aggregates.

Markedly increased heterogeneity in the cell surface properties of the amoebae was first detected by TLCCD at 8 h development and, to determine how this was related to cell differentiation, amoebae taken from either end of the 8 h profile (*Figure 5*) were mixed and allowed to form fruiting bodies. As at 10 h development, it was found that the amoebae with the more hydrophilic cell surface properties all formed spores and that stalk cells were formed from the amoebae with the more hydrophobic cell surface properties (Sharpe *et al.*, 1982). Of the eight polypeptides gained specifically by the presumptive spore cells by 10 h development, three had also been gained specifically by the presumptive spore cells by 8 h development (Sharpe *et al.*, 1985). Similarly, three polypeptides were identified that had been lost, both at 8 h and 10 h development, only by the presumptive spore cells. By contrast, no changes in polypeptide composition were detected at 8 h development that were unique to the presumptive stalk cells. This may imply that spore cell, but *not* stalk cell, differentiation is initiated by 8 h development and that, at this time of development when cell aggregation begins, the population of amoebae probably comprises undifferentiated amoebae and amoebae following the spore pathway of differentiation. One of the polypeptides unique to the presumptive spore cells at both 8 h and 10 h development was also found in the amoebae with the more hydrophilic, cell-surface properties at 6 h development (P.T.Sharpe, personal communication) and it is therefore probable that spore differentiation is proceeding even at this very early stage of development. This conclusion has been further supported by the electronmicroscopic studies (Schaap, 1983). A different conclusion could be reached on the basis of the experiments described by Tasaka *et al.* (1983) who made use of a monoclonal antibody that recognizes an antigen present only in the prestalk cells of migrating slugs. The antigen is not present in amoebae during growth but can be found in an increasing proportion of the amoebae after 3 h development and this could indicate that stalk cell differentiation is initiated very early in development and before spore cell differentiation. However, it is not clear whether this antigen should be used to detect amoebae following the stalk pathway of differentiation during early development because the antigen also appears in amoebae that will eventually become prespore cells and from which the antigen is eventually lost.

If spore differentiation is initiated prior to stalk differentiation, a further, possible explanation can be proposed for the paradoxical distribution of the stalk-inducing morphogen DIF which, as described previously, is present at a higher concentration in the prespore cells than in the prestalk cells of a migrating slug. Since DIF synthesis does not start until 9–10 h development (Brookman *et al.*, 1982), whereas spore differentiation is probably initiated a few hours earlier, it is possible that DIF is synthesized and secreted only by the presumptive spore cells. [Furthermore, it must be synthesized by the prespore cells in the *in vitro* system for studying development (Kay *et al.*, 1978).] DIF would therefore be at highest concentration among the prespore cells but could, after secretion, have its effects only on undifferentiated amoebae and induce them to follow the stalk pathway of differentiation. Presumbly an early event in the presumptive stalk cells must be loss of the inhibitors of cyclic AMP phosphodiesterase and 5′-nucleotidase which will lead to these two enzymes having higher activities in the prestalk cells than in the prespore cells of a migrating slug. As a result, a concentration gradient of adenosine could be created along the anterior–posterior axis of the slug.

Cell sorting out

There is now considerable evidence suggesting that the relationship between cell differentiation and aggregate polarity in *D. discoideum* arises as a result of sorting out of the spore and stalk precursor cells to create what will eventually become the prestalk-prespore pattern which correlates with the anterior–posterior polarity of a migrating slug. However, at present, the mechanisms controlling cell sorting are poorly understood despite the great importance of the process.

Most often, it has been proposed that cell sorting is based on differential cell cohesion (Forman and Garrod, 1977b; Tasaka and Takeuchi, 1979; Feinberg *et al.*, 1979; Lam *et al.*, 1981; Sharpe *et al.*, 1982; Siu *et al.*, 1983). This proposal is based on a general theory described by Steinberg (1964) which, if applied to *D. discoideum* amoebae, would require that the strength of intercellular cohesion between the stalk precursor cells should be different from that between the spore precursor cells. Sorting out of the two cell types would then occur to give the energetically most favourable cell arrangement but the actual arrangement (pattern) would depend on the strength of cohesion between cells of the two, different types. Cohesion between a spore precursor cell and a stalk precursor cell would have to be weaker than the cohesion between two spore precursor cells, and also weaker than the cohesion between two stalk precursor cells, for it to be expected that cell sorting out would give a pattern similar to that formed in a migrating slug or in the spherical aggregates formed when amoebae are maintained in shaken culture. At present, it is not known whether the cohesive properties of differentiating amoebae are consistent with these requirements.

In normal conditions of development, cell sorting appears to be very rapid (Tasaka and Takeuchi, 1981; Schaap *et al.*, 1982; Krefft *et al.*, 1984; Noce and Takeuchi, 1985) and to occur as the hemispherical aggregates gain apical tips (11 h development – *Figure 1*). It is therefore important that the cohesive properties of the two, differentiating cell types at this stage of development should be investigated. Nevertheless, this still remains to be done although Sharpe *et al.* (1982) have claimed that, at 10 h development, the presumptive spore cells may be more cohesive than the presumptive stalk cells. There have also been a number of studies of cell cohesion that have made use of prestalk and prespore cells separated from migrating slugs (Tasaka and Takeuchi, 1979; Lam *et al.*, 1981; Tsang and Bradbury, 1981) although, at this stage of development, the cells may have cohesive properties different from those displayed by cells at the stage of development where sorting out and pattern formation occur. Although all the studies have indicated that the prespore cells differ from the prestalk cells in their cohesive properties, there has been disagreement concerning which cell type is the more cohesive. Furthermore, Tasaka and Takeuchi (1979) have found that there are changes in the relative strengths of cohesion of the two cell types during the course of an experiment. Experiments making use of separated prespore and prestalk cells from migrating slugs have also implicated a plasma membrane glycoprotein (gp 150) in cell sorting because Fab fragments directed against this glycoprotein prevented cell sorting in an experimental system, whereas Fab fragments directed against other plasma membrane glycoproteins were without effect (Siu *et al.*, 1983). Since gp 150 is present in amoebae even during early development, it could be involved in the cell sorting that occurs in early aggregates. The role of gp 150 is still unclear, however, because at the time when cell sorting occurs it is not present in the spore and stalk precursor cells in different amounts (Geltosky *et al.*, 1979). Another approach is to try and identify plasma membrane glycoproteins that are present in one differentiating cell type, but not the other, because cells may have different sorting properties because of the presence, or absence, of these proteins in their plasma membranes. In preliminary experiments, presumptive spore and stalk cells were separated by partitioning at 10–11 h development and the plasma membrane proteins of the intact cells were labelled with ^{125}I in the reaction catalysed by lactoperoxidase. After the cell proteins had been separated by polyacrylamide-gel electrophoresis, it was possible to identify a small number of ^{125}I-labelled, plasma membrane proteins that were cell-type-specific (D.J.Watts, unpublished results).

Cell sorting has also been shown to occur in clumps of prespore and prestalk cells (derived from migrating slugs) during chemotaxis of the cells towards cyclic AMP. This occurs because only the prestalk cells are chemotactic to cyclic AMP (Matsukuma and Durston, 1979; Sternfeld and David, 1981; Mee *et al.*, 1986). This could indicate that cell sorting during normal development is also dependent on differential chemotaxis but, for this to occur, there would have to be a concentration gradient of cyclic AMP in a hemispherical aggregate at 10–11 h development. A suitable concentration gradient, with the highest concentration of cyclic AMP at the apex of the aggregate, might be created owing to diffusion of cyclic AMP out of the basal region of the aggregate into the substratum (Matsukuma and Durston, 1979; Sternfeld and David, 1981), but, at present, there is little evidence to support this view. [Studies of control of mRNA accumulation in the cells of hemispherical aggregates would suggest that there must be high, extracellular concentrations of cyclic AMP throughout the aggregates (Williams *et al.*, 1980; Chung *et al.*, 1981).] Presumably, the pattern formation observed in aggregates formed during development of amoebae in shaken culture cannot be dependent on differential chemotaxis since it would appear impossible for suitable concentration gradients of cyclic AMP to be created in these aggregates. However, cell sorting in these aggregates, which is slower than in the hemispherical aggregates formed during normal development on a solid substratum, could be entirely dependent on differential cell cohesion whereas, as proposed by Oshmori and Maeda (1986), the more rapid cell sorting in hemispherical aggregates could be dependent on both differential cohesiveness and differential chemotaxis.

Morphological polarity and cell sorting

Aggregate polarity is first recognizable when a hemispherical aggregate gains an apical tip which then persists throughout the remaining stages of development. The tip merges into the prestalk region of an aggregate and, although it is not clear to what extent the cells of the tip should be considered as being different from prestalk cells (see MacWilliams and David, 1984), it is highly probable that formation of the tip is dependent on the cell sorting that also creates the prestalk–prespore pattern. This is most obvious in the atypical conditions created by removing the anterior, prestalk region from the posterior, prespore region of a migrating slug. Although the posterior part of the slug predominantly comprises prespore cells, it also contains some (15%) anterior-like cells (Sternfeld and David, 1981, 1982; Devine and Loomis, 1985; Kakutani and

Takeuchi, 1986), which have many properties in common with prestalk cells and, when the prestalk region has been removed from a slug, these cells sort out from the prespore cells and migrate to the cut surface of the posterior region to create both a new anterior tip and a new prestalk zone. Formation of the tip can also be followed in aggregates formed from mixtures of amoebae that have been grown in axenic culture medium containing glucose [G(+) cells], or lacking glucose [G(−) cells]. The G(+) cells and the G(−) cells can then be identified within each aggregate; it is now well established that the cells sort out and that, eventually, the G(+) cells form spores whereas the G(−) cells form stalk cells (Leach *et al.*, 1973; Takeuchi *et al.*, 1982). Sorting out of the stalk precursor cells [G(−)] into the apex of a hemispherical aggregate appears to occur simultaneously with formation of the apical tip (Tasaka and Takeuchi, 1981; Noce and Takeuchi, 1985) and this is consistent with the conclusion that the tip and the prestalk region of an aggregate are both created as a result of cell sorting. Hence pattern formation and formation of a polar aggregate both seem to be the result of cell sorting and are therefore dependent on prior cell differentiation.

Control of the proportions of prespore and prestalk cells

The hypothesis that cell differentiation in *D. discoideum* might be controlled by positional information at first appeared helpful because it could explain not only how pattern formation is controlled but also how the correct proportions of prestalk and prespore cells are established (*Figure 4*). The alternative hypothesis, namely that cell differentiation is initiated early in development and gives a random mixture of differentiating cells, which then sort out is supported by considerable experimental evidence. Although it may provide a sufficient explanation for pattern formation, it offers no explanation for control of cell proportioning. It seems probable, at present, that cell proportioning is controlled by feedback mechanisms based on chemical signalling between the differentiating cells (MacWilliams and David, 1984). By using suspensions of prestalk and prespore cells, it has been possible to investigate cell proportioning (Sakai, 1973; Tasaka and Takeuchi, 1979; Takeuchi *et al.*, 1982; Weijer and Durston, 1985) and to show that it is affected by the *D. discoideum* morphogens cyclic AMP and adenosine. These experiments, together with those showing that pattern formation occurs in aggregates formed during development of amoebae in shaken culture, imply that cell proportioning need not be controlled by positional information. However, there are a number of experiments that indicate that cell proportioning can respond to positional effects.

Positional effects on cell proportioning are particularly evident during development of a number of species of cellular slime moulds related to *D. discoideum*. For example, in *Dictyostelium minutum*, aggregates are formed in which all the amoebae first differentiate into prespore cells but, later, the prespore cells at the apex of the aggregate redifferentiate into stalk cells (Schaap *et al.*, 1983). Similar, position-dependent, cell redifferentiation also occurs during development of *Dictyostelium mucoroides* (Bonner *et al.*, 1955; Gregg and Davis, 1983) and *Polysphondylium* (Hohl *et al.*, 1977; O'Day, 1979). Position-dependent effects on differentiation can also be observed in *D. discoideum*, particularly in abnormal conditions of development. When the prestalk region of a migrating slug is separated from the prespore region, each part of the slug can complete development and give rise to a fruiting body (Raper, 1940; Bonner, 1952;

Sampson, 1976). This occurs because the prestalk cells at the posterior of the prestalk section redifferentiate into prespore cells (Takeuchi *et al.*, 1982) whereas, in the prespore section, a new anterior prestalk region is created owing to migration of anterior-like cells to the anterior and to redifferentiation of prespore cells into prestalk cells (Sternfeld and David, 1982; Takeuchi *et al.*, 1982).

One hypothesis that could account for both cell sorting and position-dependent effects being involved in cell proportioning and pattern formation in *D. discoideum* is dependent on evidence (Matsukuma and Durston, 1979; Sternfeld and David, 1981; Sharpe *et al.*, 1982; Schaap *et al.*, 1982; Schaap, 1983; Krefft *et al.*, 1985) that the presumptive stalk cells have properties little different from those of undifferentiated amoebae. Cell sorting to create polar aggregates may therefore essentially involve prespore cells and undifferentiated amoebae rather than prespore cells and prestalk cells. After cell sorting has been completed, undifferentiated amoebae might be induced to follow the stalk pathway of differentiation in response to stalk-inducing morphogens, such as adenosine and DIF, and this effect would clearly be position-dependent because the cells responding to the morphogens would, as a result of cell sorting, all be positioned at the apex of the aggregate. However, it should not be assumed that all of the undifferentiated amoebae at the apex of the aggregate eventually differentiate into stalk cells. Results of several investigations indicate that the population of amoebae in a hemispherical aggregate, before cell sorting has started, contains a smaller proportion of prespore cells than the population of amoebae in a migrating slug after cell sorting has been completed (Hayashi and Takeuchi, 1976; Sharpe *et al.*, 1982; Schaap *et al.*, 1982; Schaap, 1983). Presumably some of the undifferentiated amoebae must differentiate into prespore cells after cell sorting out has occurred. Thus, after cell sorting has taken place, there could be several position-dependent events that have to occur in order that the position of the boundary between the prestalk and prespore regions of the slug can be fixed, and these events would be the 'fine control' over cell proportioning.

The theory would also account for the cell sorting that occurs in spherical aggregates formed from amoebae maintained in shaken culture and in which, it is assumed, events depending on aggregate polarity cannot occur. The theory might also appear to predict that, in these aggregates, those amoebae that do not differentiate into prespore cells should make little progress along the pathway leading to stalk cell differentiation. However, as mentioned earlier, the studies of Weijer and Durston (1985) could indicate that *D. discoideum* morphogens, such as adenosine and cyclic AMP, may be able to control cell proportioning in a manner that does not necessarily depend on cell position within a polar aggregate.

Finally, if as suggested by Sharpe *et al.* (1982) and Sharpe and Watts (1985a), cell sorting is dependent on the differences in cell surface properties that are detected by TLCCD, it is also possible to explain why, in aggregates formed from mixtures of G(+) and G(−) amoebae, the cells sort out so that the G(+) cells form spores and the G(−) cells form stalk cells. In aggregates containing only G(+) cells, it is the cells with the more hydrophobic cell surface properties that sort to the anterior of a migrating slug, whereas the cells with more hydrophilic cell surface properties sort to the posterior. Similarly, in mixtures of G(+) and G(−) cells, it is the cells with the more hydrophobic surface properties [G(−) cells] that sort to the anterior of the slug (Sharpe and Watts, 1985b).

Conclusion and summary

Extensive studies on development and differentiation in cellular slime moulds have given apparently conflicting results concerning the relationship between cell polarity and cell fate. There is considerable evidence that cell differentiation and pattern formation can occur in *D. discoideum* without any need for polar aggregates to be formed but, on the other hand, there is also evidence which equally as strongly indicates that, in *D. discoideum* and related species of the cellular slime moulds, cell differentiation and pattern formation may be controlled by position-dependent events that could occur only in polar aggregates. The diffficulties arise primarily out of investigations of the very early stages of cell differentiation and it is probable that they would be resolved if a means were available for identifying unequivocally the progress of spore and stalk cell differentiation during early development. A collection of monoclonal antibodies suitable for this purpose would be invaluable. A better understanding of the control of initiation of cell differentiation (see Sharpe and Watts, 1985a) would also be helpful. Meanwhile, it is to be expected that continuation of studies of the functions of the cellular slime mould morphogens will allow further progress to be made towards an understanding of the extent to which cell differentiation is controlled by positional information. In many respects, it is the remarkable success in identifying these morphogens that makes the cellular slime moulds still extremely attractive as systems in which to study control of cell differentiation.

Acknowledgements

I am grateful to Dr D.R.Garrod, Dr T.E.Treffry and Mr P.Hancock for their help with the illustrations and to Dr J.J.Brookman, Ms. G.M.Knight and Dr P.T.Sharpe for allowing me to refer to unpublished results.

References

Alton,T.H. and Brenner,M. (1979) Comparison of proteins synthesized by anterior and posterior regions of *Dictyostelium discoideum* pseudoplasmodia. *Developmental Biology,* **71**, 1–7.

Armant,D.R. and Rutherford,C.L. (1979) 5′-AMP nucleotidase is localized in the area of cell–cell contact of prespore and prestalk regions during culmination of *Dictyostelium discoideum. Mechanisms of Ageing and Development,* **10**, 199–217.

Armant,D.R. and Rutherford,C.L. (1981) Copurification of alkaline phosphatase and 5′-AMP specific nucleotidase in *Dictyostelium discoideum. Journal of Biological Chemistry,* **256**, 12710–12718.

Armant,D.R., Stetler,D.A. and Rutherford,C.L. (1980) Cell surface localization of 5′-AMP nucleotidase in prestalk cells of *Dictyostelium discoideum. Journal of Cell Science,* **45**, 119–129.

Barklis,E. and Lodish,H.F. (1983) Regulation of *Dictyostelium discoideum* mRNAs specific for prespore or prestalk cells. *Cell,* **32**, 1139–1148.

Bonner,J.T. (1952) The pattern of differentiation in amoeboid slime molds. *American Naturalist,* **86**, 79–89.

Bonner,J.T. (1970) Induction of stalk cell differentiation by cyclic AMP in the cellular slime mold *Dictyostelium discoideum. Proceedings of the National Academy of Science USA,* **72**, 3163–3166.

Bonner,J.T., Chiqoine,A.D. and Kolderie,M.J. (1955) A histochemical study of differentiation in the cellular slime molds. *Journal of Experimental Zoology,* **130**, 133–157.

Brackenbury,R.N., Schindler,J., Alexander,S. and Sussman,M. (1974) A choice of morphogenetic pathways in *Dictyostelium discoideum* induced by the adenosine analog formycin B. *Journal of Molecular Biology,* **90**, 529–539.

Brenner,M. (1977) Cyclic AMP gradient in migrating pseudoplasmodia of the cellular slime mold *Dictyostelium discoideum. Journal of Biological Chemistry,* **252**, 4073–4077.

Brookman,J.J., Town,C.D., Jermyn,K.A. and Kay,R.R. (1982) Developmental regulation of a stalk cell differentiation-inducing factor in *Dictyostelium discoideum. Developmental Biology,* **91**, 191–196.

Brown,S.S. and Rutherford,C.L. (1980) Localization of cyclic nucleotide phosphodiesterase in multicellular stages of *Dictyostelium discoideum. Differentiation,* **16**, 173–183.

Chung,S., Landfear,S.M., Blumberg,D.D., Cohen,N.S. and Lodish,H.F. (1981) Synthesis and stability of developmentally regulated *Dictyostelium* mRNAs are affected by cell–cell contact and cyclic AMP. *Cell,* **24**, 785–797.

Das,D.V.M. and Weeks,G. (1980) Reversible heat activation of alkaline phosphatase of *Dictyostelium discoideum* and its developmental implication. *Nature,* **288**, 166–167.

Das,D.V.M. and Weeks,G. (1981) The inhibition of *Dictyostelium discoideum* alkaline phosphatase by a low molecular weight factor and its implications for the developmental regulation of the enzyme. *FEBS Letters,* **130**, 249–252.

Datta,S., Gomer,R.H. and Firtel,R.A. (1986) Spatial and temporal regulation of a foreign gene by a prestalk-specific promoter in transformed *Dictyostelium discoideum. Molecular and Cellular Biology,* **6**, 811–820.

Devine,K.M .and Loomis,W.F. (1985) Molecular characterisation of anterior-like cells in *Dictyostelium discoideum. Developmental Biology,* **107**, 364–372.

Devine,K.M., Bergmann,J.E. and Loomis,W.F. (1983) Spore coat proteins of *Dictyostelium discoideum* are packaged in prespore vacuoles. *Developmental Biology,* **99**, 437–446.

Dominov,J.A. and Town,C.D. (1986) Regulation of stalk and spore antigen expression in monolayer cultures of *Dictyostelium discoideum* by pH. *Journal of Embryology and Experimental Morphology,* **96**, 131–150.

Feinberg,A.P., Springer,W.R. and Barondes,S.H. (1979) Segregation of pre-stalk and pre-spore cells of *Dictyostelium discoideum*: Observations consistent wtih selective cell cohesion. *Proceedings of the National Academy of Science USA,* **79**, 3977–3981.

Forman,D. and Garrod,D.R. (1977a) Pattern formation in *Dictyostelium discoideum.* I. Development of prespore cells and its relationship to the pattern of the fruiting body. *Journal of Embryology and Experimental Morphology,* **40**, 215–228.

Forman,D. and Garrod,D.R. (1977b) Pattern formation in *Dictyostelium discoideum.* II. Differentiation and pattern formation in non-polar aggregates. *Journal of Embryology and Experimental Morphology,* **40**, 229–243.

Garrod,D.R. and Malkinson,A.M. (1973) Cyclic AMP, pattern formation and movement in the slime mould *Dictyostelium discoideum. Experimental Cell Research,* **81**, 492–495.

Geltosky,J.E., Weseman,J., Bakke,A. and Lerner,R.A. (1979) Identification of a cell surface glycoprotein involved in cell aggregation in *D. discoideum. Cell,* **18**, 391–396.

George,R.P., Hohl,H.R. and Raper,K.B. (1972) Ultrastructural development of stalk producing cells in *Dictyostelium discoideum,* a cellular slime mould. *Journal of General Microbiology,* **40**, 477–489.

Gerisch,G., Malchow,D., Riedel,V., Muller,E. and Every,M. (1972) Cyclic AMP phosphodiesterase and its inhibitor in slime mould development. *Nature, New Biology,* **235**, 90–92.

Gierer,A. and Meinhardt,H. (1972) A theory of biological pattern formation. *Kybernetik,* **12**, 30–39.

Gregg,J.H. and Badman,W.S. (1973) Morphogenesis and ultrastructure in *Dictyostelium. Developmental Biology,* **22**, 96–111.

Gregg,J.H. and Karp,G.C. (1978) Patterns of cell differentiation revealed by L-[^{3}H]fucose incorporation in *Dictyostelium. Experimental Cell Research,* **112**, 31–46.

Gregg,J.H. and Davis,R.W. (1983) Dynamics of cell redifferentiation in *Dictyostelium mucoroides. Differentiation,* **21**, 200–205.

Gross,J.D., Town,C.D., Brookman,J.J., Jermyn,K.A., Peacey,M.J. and Kay,R.R. (1981) Cell patterning in *Dictyostelium. Philosophical Transactions of the Royal Society London B,* **295**, 497–508.

Hayashi,M. and Takeuchi,I. (1976) Quantitative studies on cell differentiation during morphogenesis of the cellular slime mold *Dictyostelium discoideum. Developmental Biology,* **50**, 302–309.

Hohl,H.R. and Hamamoto,S.T. (1969) Ultrastructure of spore differentiation in *Dictyostelium*: the prespore vacuole. *Journal of Ultrastructural Research,* **26**, 442–453.

Hohl,H.R., Honegger,R., Traub,F. and Markwalder,M. (1977) Influence of cAMP on cell differentiation and morphogenesis in *Polysphondylium.* In *Development Growth and Differentiation in Cellular Slime Moulds.* Cappuccinelli,P. and Ashworth,J.M. (eds), Elsevier/North Holland Biomedical Press, Amsterdam, pp. 149–172.

Kakutani,T. and Takeuchi,I. (1986) Characterisation of anterior-like cells in *Dictyostelium* as analyzed by their movement. *Developmental Biology,* **115**, 439–445.

Kay,R.R. (1979) Gene expression in *Dictyostelium discoideum*: mutually antagonistic roles of cyclic-AMP and ammonia. *Journal of Embryology and Experimental Morphology,* **52**, 171–182.
Kay,R.R. (1982) cAMP and spore differentiation in *Dictyostelium discoideum. Proceedings of the National Academy of Science USA,* **79**, 3228–3231.
Kay,R.R. and Jermyn,K.A. (1983) A possible morphogen controlling differentiation in *Dictyostelium. Nature,* **303**, 242–244.
Kay,R.R., Garrod,D.R. and Tilly,R. (1978) Requirements for cell differentiation in *Dictyostelium discoideum. Nature,* **271**, 58–60.
Kay,R.R., Dhokia,B. and Jermyn,K.A. (1983) Purification of stalk-cell-inducing morphogens from *Dictyostelium discoideum. Euorpean Journal of Biochemistry,* **136**, 51–56.
Konijn,T.M., Van de Meene,J.G.C., Bonner,J.T. and Barkley,D.S. (1967) The acrasin activity of adenosine-3′,5′-cyclic phosphate. *Proceedings of the National Academy of Science USA,* **58**, 1152–1154.
Kopachik,W., Oohata,A., Dhokia,B., Brookman,J.J. and Kay,R.R. (1983) *Dictyostelium* mutants lacking DIF, a putative morphogen. *Cell,* **33**, 397–403.
Kopachik,W., Dhokia,B. and Kay,R.R. (1985) Selective induction of stalk-cell-specific proteins in *Dictyostelium. Differentiation*, **28**, 209–216.
Krefft,M., Voet,L., Gregg,J.H., Mairhofer,H. and Williams,K.L. (1984) Evidence that positional information is used to establish the pre-stalk-pre-spore pattern in *Dictyostelium discoideum. The EMBO Journal,* **3**, 201–206.
Krefft,M., Voet,L., Gregg,J.H. and Williams,K.L. (1985) Use of a monoclonal antibody recognizing a cell surface determinant to distinguish prestalk and prespore cells in *Dictyostelium discoideum* slugs. *Journal of Embryology and Experimental Morphology,* **88**, 15–24.
Kumagai,A. and Okamoto,K. (1986) Prespore-inducing factors in *Dictyostelium discoideum. Differentiation*, **31**, 79–84.
Lam,T.Y., Pickering,G., Geltosky,J. and Siu,C.-H. (1981) Differential cell cohesiveness expressed by prespore and prestalk cells of *Dictyostelium discoideum. Differentiation,* **20**, 22–28.
Leach,C.K., Ashworth,J.M. and Garrod,D.R. (1973) Cell sorting out during the differentiation of mixtures of metabolically distinct populations of *Dictyostelium discoideum. Journal of Embryology and Experimental Morphology,* **29**, 647–661.
MacWilliams,H.K. and Bonner,J.T. (1979) The prestalk–prespore pattern in cellular slime molds. *Differentiation,* **14**, 1–22.
MacWilliams,H.K. and David,C.N. (1984) Pattern formation in *Dictyostelium*. In *Microbial Development*. Losick,R. and Shapiro,L. (eds), *Cold Spring Harbor Monograph, Volume 16,* Cold Spring Harbor Laboratory Press, Cold Spring Harbor, NY, pp. 255–274.
Matsukuma,S. and Durston,A.J. (1979) Chemotactic cell sorting in *Dictyostelium discoideum. Journal of Embryology and Experimental Morphology,* **50**, 243–251.
Mee,J.D., Tortolo,D.M .and Coukell,M.B. (1986) Chemotaxis-associated properties of separated prestalk and prespore cells of *Dictyostelium discoideum. Biochemistry and Cell Biology,* **64**, 722–732.
Mehdy,M.C., Ratner,D. and Firtel,R.A. (1983) Induction and modulation of cell-type-specific gene expression in *Dictyostelium. Cell,* **32**, 763–771.
Meinhardt,H. (1983) A model for the pre-stalk/pre-spore patterning in the slug of the slime mould *Dictyostelium discoideum. Differentiation,* **24**, 191–202.
Morrissey,J.H., Devine,K.M. and Loomis,W.F. (1984) The timing of cell-type specific differentiation in *Dictyostelium discoideum. Developmental Biology,* **103**, 414–424.
Neave,N., Sobsolewski,A. and Weeks,G. (1983) The effect of ammonia on stalk cell formation in submerged monolayers of *Dictyostelium discoideum. Cell Differentiation,* **13**, 301–307.
Neave,N., Kwong,L., MacDonald,J.I.S. and Weeks,G. (1986) The distribution of stalk cell differentiation inducing factor and other lipids during the differentiation of *Dictyostelium discoideum. Biochemistry and Cell Biology,* **64**, 85–90.
Newell,P.C. (1982) Cell surface binding of adenosine to *Dictyostelium* and inhibition of pulsatile signalling. *FEMS Microbiology Letters,* **13**, 417–421.
Newell,P.C. and Ross,F.M. (1982) Inhibition by adenosine of aggregation centre initiation and cyclic AMP binding in *Dictyostelium. Journal of General Microbiology*, **128**, 2715–2724.
Newell,P.C., Ellingson,J.S. and Sussman,M. (1969) Synchrony of enzyme accumulation in a population of differentiating slime mold cells. *Biochimica et Biophysica Acta,* **177**, 610–614.
Noce,T. and Takeuchi,I. (1985) Prestalk/prespore differentiation tendency of *Dictyostelium discoideum* cells

as detected by a stalk-specific monoclonal antibody. *Developmental Biology*, **109**, 157–164.
O'Day,D.H. (1979) Cell differentiation during fruiting body formation in *Polysphondylium pallidum. Journal of Cell Science*, **35**, 203–215.
Oshmori,T. and Maeda,Y. (1986) Implications of differential chemotaxis and cohesiveness for cell sorting in the development of *Dictyostelium discoideum. Development, Growth and Differentiation*, **28**, 169–175.
Pan,P., Bonner,J.T., Wedner,H.J. and Parker,C.W. (1974) Immunofluorescence evidence for the distribution of cyclic AMP in cells and cell masses of the cellular slime molds. *Proceedings of the National Academy of Science USA*, **71**, 1623–1625.
Raper,K.B. (1940) Pseudoplasmodium formation and organization in *Dictyostelium discoideum. Journal of the Elisha Mitchell Scientific Society*, **56**, 241–282.
Ratner,D. and Borth,W. (1983) Comparison of differentiating *Dictyostelium discoideum* cell types separated by an improved method of density gradient centrifugation. *Experimental Cell Research*, **143**, 1–13.
Rossomando,E.F. and Cutler,L.S. (1975) Localization of adenylate cyclase in *Dictyostelium discoideum. Experimental Cell Research*, **95**, 67–78.
Sakai,Y. (1973) Cell type conversion in isolated prestalk and prespore fragments of the cellular slime mold *Dictyostelium discoideum. Development, Growth and Differentiation*, **13**, 231–240.
Sampson,J. (1976) Cell patterning in migrating slugs of *Dictyostelium discoideum. Journal of Embryology and Experimental Morphology*, **36**, 663–668.
Schaap,P. (1983) Quantitative analysis of the spatial distribution of ultrastructural differentiation markers during development of *Dictyostelium discoideum. Roux's Archives of Developmental Biology*, **192**, 86–94.
Schaap,P. and Spek,W. (1984) cAMP binding to the cell surface during development of *Dictyostelium discoideum. Differentiation*, **27**, 83–87.
Schaap,P. and Wang,M. (1984) The possible involvement of oscillatory cAMP signalling in multicellular morphogenesis of the cellular slime molds. *Developmental Biology*, **105**, 470–478.
Schaap,P. and Wang,M. (1986) Correlations between tip dominance, prestalk/prespore pattern, and cyclic AMP-relay efficiency in slugs of *Dictyostelium discoideum. Differentiation*, **30**, 7–14.
Schaap,P., Van der Molen,L. and Konijn,T.M. (1982) Early recognition of prespore differentiation in *Dictyostelium discoideum* and its significance for models of pattern formation. *Differentiation*, **22**, 1–5.
Schaap,P., Van der Molen,L. and Konijn,T.M. (1983) The organisation of fruiting body formation in *Dictyostelium minutum. Cell Differentiation*, **12**, 287–297.
Schindler,J. and Sussman,M. (1977) Ammonia determines the choice of morphogenetic pathways in *Dictyostelium discoideum. Journal of Molecular Biology*, **116**, 161–169.
Schindler,J. and Sussman,M. (1979) Inhibition by ammonia of intracellular cAMP accumulation in *Dictyostelium discoideum*: Its significance for the regulation of morphogenesis. *Developmental Genetics*, **1**, 13–20.
Sharpe,P.T. (1984) Cell surface 'chromatography'. *Trends in Biochemical Sciences*, **9**, 374–377.
Sharpe,P.T. and Watts,D.J. (1984) Cell cycle-related changes in the surface properties of amoebae of the cellular slime mould *Dictyostelium discoideum. FEBS Letters*, **168**, 89–92.
Sharpe,P.T. and Watts,D.J. (1985a) The role of the cell cycle in differentiation of the cellular slime mould *Dictyostelium discoideum. Molecular and Cellular Biochemistry*, **67**, 3–9.
Sharpe,P.T. and Watts,D.J .(1985b) Use of aqueous two-phase partition to detect cell surface changes during growth of *Dictyostelium discoideum. Journal of Cell Science*, **73**, 339–346.
Sharpe,P.T., Treffry,T.E. and Watts,D.J. (1982) Studies of early stages of differentiation of the cellular slime mould *Dictyostelium discoideum. Journal of Embryology and Experimental Morphology*, **67**, 181–193.
Sharpe,P.T., Knight,G.M .and Watts,D.J. (1984) Changes in the DNA content of amoebae of *Dictyostelium discoideum* during growth and development. *Biochemical Journal*, **217**, 839–843.
Sharpe,P.T., Sharrard,R.M. and Watts,D.J. (1985) Polypeptide compositions of amoebae of the cellular slime mould *Dictyostelium discoideum* separated by partitioning during development. *Bioscience Reports*, **5**, 121–127.
Siu,C.-H., Des Roches,B. and Lam,T.Y. (1983) Involvement of a cell-surface glycoprotein in the cell-sorting process of *Dictyostelium discoideum. Proceedings of the National Academy of Science USA*, **80**, 6596–6600.
Sobsoleweski,A., Neave,N. and Weeks,G. (1983) The induction of stalk cell differentiation in submerged monolayers of *Dictyostelium discoideum*. Characterisation of the temporal sequence of molecular requirements. *Differentiation*, **25**, 93–100.
Steinberg,M.S. (1964) The problem of adhesive selectivity in cellular interactions. In *Cellular Membranes in Development*. Locke,M. (ed.), Academic Press, New York, pp. 321–366.
Sternfeld,J. and David,C.N. (1981) Cell sorting during pattern formation in *Dictyostelium. Differentiation*, **20**, 10–21.

Sternfeld,J. and David,C.N. (1982) Fate and regulation of anterior-like cells in *Dictyostelium* slugs. *Developmental Biology,* **93**, 111–118.

Sussman,M. and Lovgren,N. (1965) Preferential release of the enzyme UDP-galactose polysaccharide transferase during cellular differentiation in the slime mold, *Dictyostelium discoideum. Experimental Cell Research,* **38**, 97–105.

Sussman,M. and Schindler,J. (1978) A possible mechanism of morphogenetic regulation in *Dictyostelium. Differentiation,* **10**, 1–5.

Sussman,M., Schindler,J. and Kim,H. (1977) Towards a biochemical definition of the morphogenetic fields in *Dictyostelium discoideum.* In *Development, Growth and Differentiation in Cellular Slime Moulds.* Cappuccinelli,P. and Ashworth,J.M. (eds), Elsevier/North Holland Biomedical Press, Amsterdam, pp. 31–50.

Takeuchi,I. (1969) Establishment of polar organisation during slime mold development. In *Nucleic Acid Metabolism in Cell Differentiation and Cancer Growth.* Cowdry,E.V. and Seno,S. (eds), Pergamon Press, Oxford, pp. 297–304.

Takeuchi,I. (1972) Differentiation and dedifferentiation in cellular slime moulds. In *Aspects of Cellular and Molecular Physiology.* Hamoguchi,K. (ed.), University of Tokyo Press, Tokyo, pp. 217–236.

Takeuchi,I., Tasaka,M., Oyama,M., Yamamoto,A. and Amagai,A. (1982) Pattern formation in the development of *Dictyostelium discoideum.* In *Embryonic Development Part B.* Burger,M.M. and Weber,R. (eds), Alan R.Liss, New York, pp. 283–294.

Tasaka,M. and Takeuchi,I. (1979) Sorting out behaviour of disaggregated cells in the absence of morphogenesis in *Dictyostelium discoideum. Journal of Embryology and Experimental Morphology,* **49**, 89–102.

Tasaka,M. and Takeuchi,I. (1981) Role of cell sorting in pattern formation in *Dictyostelium discoideum. Differentiation.,* **18**, 191–196.

Tasaka,M., Noce,T. and Takeuchi,I. (1983) Prestalk and prespore differentiation in *Dictyostelium* as detected by cell type-specific monoclonal antibodies. *Proceedings of the National Academy of Science USA,* **80**, 5340–5344.

Toda,K., Ono,K. and Ochiai,H. (1980) Surface labelling of membrane glycoproteins and their drastic changes during development of *Dictyostelium discoideum. European Jounal of Biochemistry,* **111**, 377–388.

Town,C.D. and Stanford,E. (1979) An oligosaccharide-containing factor that induces cell differentiation in *Dictyostelium discoideum. Proceedings of the National Academy of Science USA,* **76**, 308–312.

Town,C.D., Gross,J.D. and Kay,R.R. (1976) Cell differentiation without morphogenesis in *Dictyostelium discoideum. Nature,* **262**, 717–719.

Treffry,T.E. and Watts,D.J. (1976) Development of *Dictyostelium discoideum*: A study by scanning electron microscopy. *Micron,* **7**, 11–20.

Tsang,A. and Bradbury,J.M. (1981) Separation and properties of prestalk and prespore cells of *Dictyostelium discoideum. Experimental Cell Research,* **132**, 433–441.

Watts,D.J. and Treffry,T.E. (1976) Culmination in the slime mould *Dictyostelium discoideum* studied with a scanning electron microscope. *Journal of Embryology and Experimental Morphology,* **35**, 323–333.

Weijer,C.J. and Durston,A.J. (1985) Influence of cyclic AMP and hydrolysis products on cell type regulation in *Dictyostelium discoideum. Journal of Embryology and Experimental Morphology,* **86**, 19–37.

Wilkinson,D.G., Wilson,J. and Hames,B.D. (1985) Spore coat protein synthesis during development of *Dictyostelium discoideum* requires a low-molecular-weight inducer and continued multicellularity. *Developmental Biology,* **107**, 38–46.

Williams,J.G., Tsang,A. and Mahbubani,H. (1980) A change in the rate of transcription of a eukaryotic gene in response to cyclic-AMP. *Proceedings of the National Academy of Science USA,* **77**, 7171–7175.

Williams,K.L., Fisher,P.R., MacWilliams,H.K. and Bonner,J.T. (1981) Cell patterning in *Dictyostelium discoideum. Differentiation,* **18**, 61–63.

Wilson,J.B. and Rutherford,C.L. (1978) ATP, trehalose and ammonium ion localization in the two cell types of *Dictyostelium discoideum. Journal of Cellular Physiology,* **94**, 37–46.

Wolpert,L. (1969) Positional information and spatial pattern of cellular differentiation. *Journal of Theoretical Biology,* **25**, 1–47.

CHAPTER 6

Is the spatial organization of fungal hyphae maintained and regulated by Woronin bodies?

P.MARKHAM, A.J.COLLINGE, J.B.HEAD and R.K.POOLE

Microbiology Department, King's College, Campden Hill Road, London W8 7AH, UK

Introduction

Woronin bodies were discovered in filamentous fungi over a century ago (Woronin, 1864) but remain relatively neglected and poorly understood organelles. Nevertheless, their appearance under the electron microscope is often quite distinctive and they have been recorded in more than fifty species (Markham and Collinge, 1987). Although several hypotheses have been advanced for the function of these organelles in vegetative hyphae and during sporogenesis, in several instances there has either been little evidence to support such ideas, or else the evidence advanced appears, upon careful examination, to be unsound. Because some of these hypotheses have begun to gain acceptance, it is appropriate in this chapter to consider the question of whether Woronin bodies play a role in maintaining and regulating spatial organization of fungal hyphae and to examine in some detail the evidence which has been used to advance such hypotheses. In order to do this it is first necessary to describe some important features of the nature and distribution of Woronin bodies.

Nature and distribution of Woronin bodies

A recent review has assembled much of the available information about Woronin bodies in vegetative fungal hyphae (Markham and Collinge, 1987). Whilst these organelles have been observed in a relatively large number of species, their occurrence amongst the true fungi (Eumycota) is restricted to two of the five major subdivisions. Woronin bodies have been observed only in members of the Ascomycotina and Deuteromycotina and not in the Basidiomycotina, Zygomycotina or Mastigomycotina. This might suggest that Woronin bodies play a dispensable role in the spatial organization of fungi but there are two major reasons why this conclusion may not be valid. Firstly, the ascomycete and deuteromycete fungi account for the great majority of all fungi, some 45 600 (72%) of approximately 63 000 known non-lichenized species (Hawksworth *et al.*, 1983). Secondly, the other fungal groups either perform septal pore plugging (the function widely assumed to be effected by these organelles) using a different but related mechanism as in the basidiomycetes, or are too simple in their construction to make the operation of this feasible, as in the case of the lower fungi (zygomycetes and mastigomycetes), which generally do not have septal pores. Additionally, there have been no reports of Woronin bodies in yeasts, even ascomycete and deuteromycete yeasts. Again it is the relatively simple structure of yeasts, compared to mycelial fungi, which may obviate the need for these organelles.

Woronin bodies were originally discovered and described using light microscopy (Woronin, 1864; Buller, 1933), but it is only by using electron microscopy that a precise description of these organelles can be obtained. A typical example is shown in *Figure 1*. Such electron microscope studies have shown Woronin bodies to be approximately circular or ovoid in cross-section and therefore presumably roughly spherical in overall shape. Measurements of these organelles have demonstrated that they range from 150 to 750 nm in diameter, with most species having Woronin bodies in the lower region of this size range. This demonstrates why there has always been considerable difficulty and confusion in studies of these organelles using light microscopy since they are very close in size to the theoretical limit of resolution of light microscopes (200 nm). In addition to the most generally observed shape of Woronin bodies, there are a few species in which apparently analogous particles have been reported to have a distinctly angular, often hexagonal shape in cross-section. These particles have been called hexagonal crystals (Shatkin and Tatum, 1959; Trinci and Collinge, 1974), but it is our belief and that of others (Trinci and Collinge, 1974) that these particles represent merely a structurally distinct sub-class of the organelles more widely termed Woronin bodies. Indeed this idea is supported by the occasional observation in species that contain the typical spherical Woronin bodies, of organelles with a more angular appearance (A.J.Collinge and J.B.Head, unpublished observations) and the observation of six-sided forms during the proposed development of Woronin bodies within microbodies (Wergin, 1973). However, Brenner and Carroll (1968) reported the observation of both Woronin bodies and hexagonal crystals in *Ascodesmis sphaerospora* and considered them not to be synonymous. A consistent feature of both spherical Woronin bodies and hexagonal crystals, is that they are bound by a single membrane. This is not apparent in every section, but is always found to be true if sufficient care is taken to obtain appropriate median sections. In general these organelles are referred to as electron-dense, but this is merely a convenient description and there is often great variability in the degree of electron-density displayed, depending on the species in which the organelles are observed, the staining procedure used, the position within the organelle at which the section is taken and possibly also the biochemical state of the organelle itself. Whilst published electron micrographs of these organelles generally show their internal matrix to be homogeneous, there have been reports that both hexagonal crystals and Woronin bodies have an internal lattice structure (Allen, 1976; Brenner and Carroll, 1968) and it is our experience that this is a consistent feature of these organelles if sections are prepared and examined with sufficient care (*Figure 1*).

Enzymic activities have been ascribed to amorphous, paracrystalline and crystalline inclusions observed within the microbodies of animals, plants and fungi (Wergin, 1973; Heath, 1977; Cole and Samson, 1979) . Wergin (1973) has suggested that Woronin bodies in *Fusarium oxysporum* may contain enzymatic activity by virtue of their

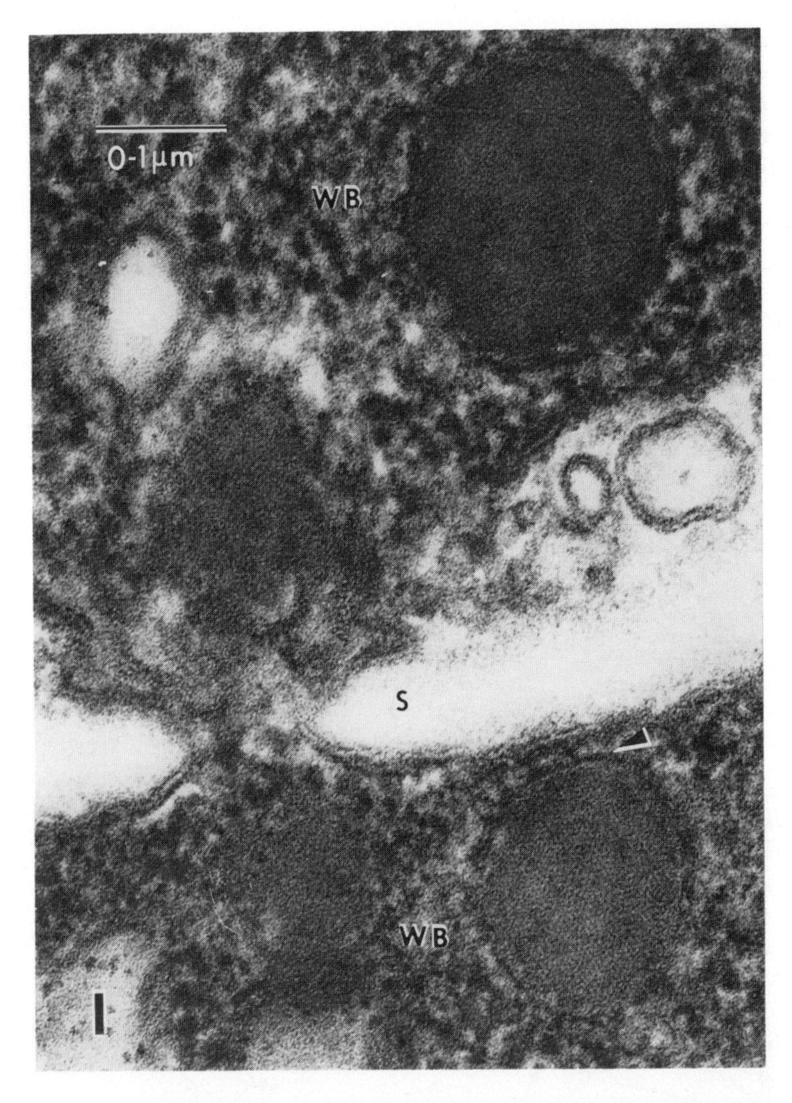

Figure 1. Transmission electron micrograph showing two typical Woronin bodies (WB) near a septum (s) in a *Penicillium chrysogenum* hypha. Note the bounding membrane (arrowed) and the regular internal lattice structure.

development within microbodies, from which they are apparently extruded, and that they may be identical to the biochemically defined lysosomes of *Ceratocystis fimbriata* (Wilson *et al.*, 1970). However, Wilson *et al.* (1970) identified as a lysosome, a structure in *Ceratocystis fagacearum* positioned close to the septal pore and showing the generally accepted structural features of a Woronin body. Woronin bodies were very poorly understood at that time and Wilson *et al.* (1970) provided no evidence linking the organelles to lysosomal enzyme activity. It therefore seems likely that the link between Woronin bodies and lysosomal enzyme activity suggested by Wergin (1973) may have been based on a confusion of Woronin bodies with other organelles of similar size.

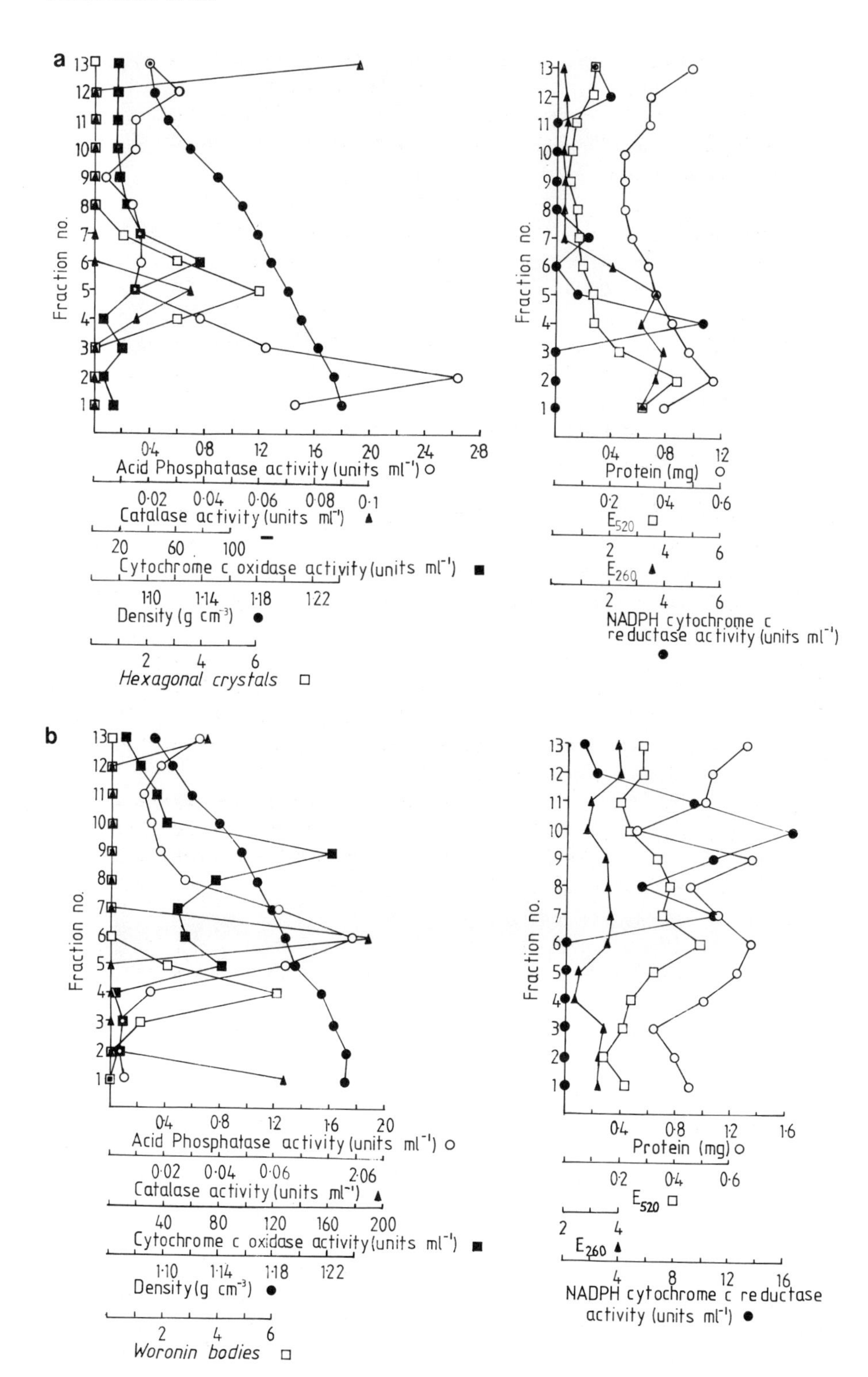
a
Fraction no.
Acid Phosphatase activity (units ml⁻¹)
Catalase activity (units ml⁻¹)
Cytochrome c oxidase activity (units ml⁻¹)
Density (g cm⁻³)
Hexagonal crystals
Protein (mg)
E520
E260
NADPH cytochrome c reductase activity (units ml⁻¹)
b
Fraction no.
Acid Phosphatase activity (units ml⁻¹)
Catalase activity (units ml⁻¹)
Cytochrome c oxidase activity (units ml⁻¹)
Density (g cm⁻³)
Woronin bodies
Protein (mg)
E520
E260
NADPH cytochrome c reductase activity (units ml⁻¹)

Allen (1976) reported the development of hexagonal crystals within microbodies of *Neurospora crassa* similar to that reported for Woronin bodies (Brenner and Carroll, 1968; Wergin, 1973). This also led to speculation about the enzymatic activity of hexagonal crystals, but Allen (1976) failed to attribute catalase activity, an enzyme commonly associated with microbodies (Vigil, 1973), to hexagonal crystals. However, Allen (1976) suggested a possible relationship between hexagonal crystals and glyoxysomes since fungal microbodies are most frequently designated as such, on the basis of their location (Mendgen, 1973), enzyme content (Powell, 1976) or relative abundance in mycelium grown on different carbon sources (Maxwell *et al.*, 1975). Glyoxysomes have indeed been isolated from *N. crassa* (Kobr *et al.*, 1969). However, the value of these reports is hampered by the confusing terminology applied to microbodies, glyoxysomes and peroxisomes (Tolbert and Essner, 1981) and the paucity of critical fractionation studies on eukaryotic microbes (Lloyd, 1974).

Analytical subcellular fractionation, introduced by Albert Claude and exploited particularly by de Duve and his collaborators (de Duve *et al.*, 1971) can provide quantitative evaluation of the distribution of a component between all fractions derived from a homogenate. It remains one of the most powerful means of correlating organelle structure and function. The procedure involves cell breakage under minimally disruptive conditions, followed by centrifugation of the resultant homogenate through a gradient of solute, designed to effect separation of organelles either on the basis of their sedimentation coefficient ('rate' separation) or buoyant density (isopycnic separation). Recently, we have performed, using high resolution zonal rotors and isopycnic separations in tubes (*Figure 2*), fractionations of homogenates from *Penicillium chrysogenum* and *N. crassa*. The peaks in the distribution pattern in isopycnic fractionations of both Woronin bodies and hexagonal crystals are distinct from the peaks of activity of marker enzymes for lysosomes (acid phosphatase) and mitochondria (cytochrome *c* oxidase). In the case of *P. chrysogenum*, the Woronin bodies were also well separated (*Figure 2b*) from peroxisomes (marker enzyme, catalase). In the fractionation of *N. crassa* homogenates, however, a peak of sedimentable catalase that was distinct from the major, soluble activity was coincident with the hexagonal crystals in fraction 5 (ϱ = 1.20 g.cm^{-3}). In this organism, glyoxysomes have been reported to have a density of 1.22 g.cm^{-3} (Kobr *et al.*, 1969) and might be expected to band in fraction 3, although glyoxysome markers were not assayed. In other microorganisms, key glyoxylate cycle enzymes and catalase are coincident in peroxisomes (e.g. Muller *et al.*, 1971; Avers, 1971; Lloyd and Cartledge, 1974). Thus the distribution of these organelles is unclear from isopycnic separations, but rate fractionations in a zonal rotor (not shown) clearly separate catalase activity from hexagonal crystals.

Evidence has been obtained from enzyme digestion and lipid solvent extraction treatments of ultrathin sections, that the bulk component of Woronin bodies is protein (McKeen, 1971; Hoch and Maxwell, 1974; Mason and Crosse, 1975). Such studies

Figure 2. Isopycnic subcellular fractionations of whole homogenates prepared from (**a**) *Neurospora crassa* and (**b**) *Penicillium chrysogenum*, on linear sucrose density gradients. The peak in the distribution of both hexagonal crystals and Woronin bodies (measured in arbitrary units based on electron microscopic observations of negatively stained fractions), are seen to be clearly separated from the peak in the distribution of all marker enzymes assayed. A unit of marker enzyme activity corresponds to 1 nmol.min^{-1}.

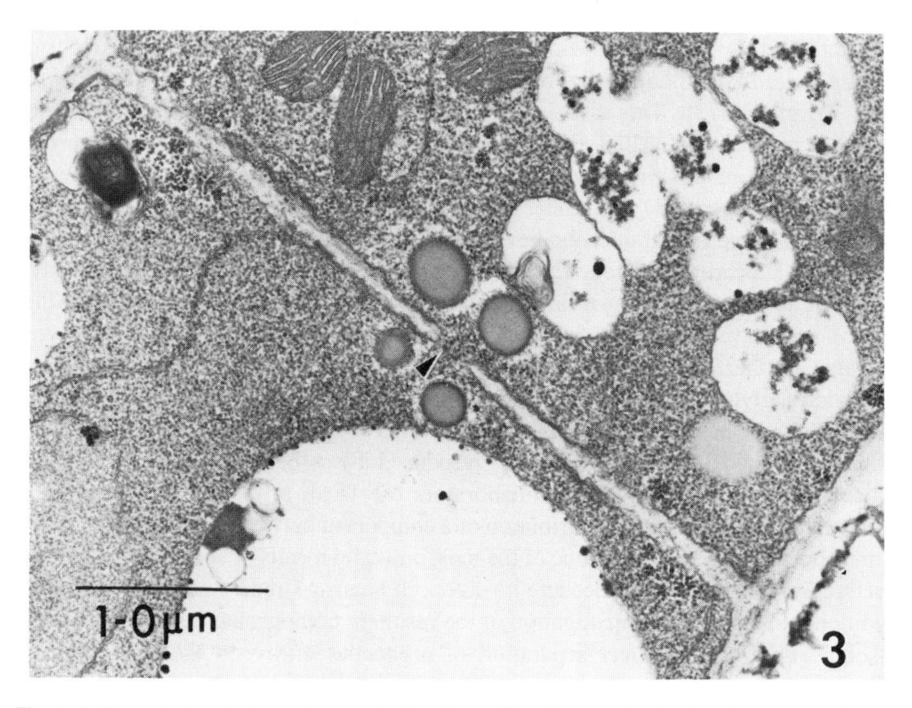

Figure 3. Transmission electron micrograph of an *Aspergillus giganteus* hypha showing a septum in which the central pore is arrowed. Four Woronin bodies can be seen in a typical arrangement, clustered close to the septal pore and on either side of the septum.

on ultrathin sections, which depend on removal of Woronin body material from sections into drops of enzyme or solvent solution are, however, suggestive rather than conclusive. This is because complete absence of the organelles from such sections might occur by loosening of the matrix material as opposed to complete dissolution. Thus, it is necessary to isolate and characterize Woronin bodies biochemically to obtain conclusive information about their composition.

It was known from the time of their discovery that Woronin bodies are distributed in a specific arrangement within fungal hyphae. Woronin (1864) reported that they were to be found close to and on either side of septa, and all subsequent observations have confirmed that this is their predominant location within hyphae. Electron microscopy has revealed that a small number of these organelles occur on each side of the cross-walls, very close to the septal plate and, clustered near to the septal pore (*Figure 3*). They are generally positioned just to one side of the pore, presumably protected by the septal plate from the flow of cytoplasm through the pore which occurs in actively growing hyphae (Buller, 1933; Gull, 1978). It is not yet known how many Woronin bodies are associated with each septum or whether the number is consistent at every septum throughout the mycelium. Certainly there are several at each septum and numbers in the range three to eight have often been reported. There is some evidence that there could be as many as twelve Woronin bodies arranged in two rings of six, one ring on each side of the septal plate (Scannerini, 1968; Collinge, 1986), as shown in *Figure 4*.

As well as a typical positional association with septal pores, Woronon bodies also

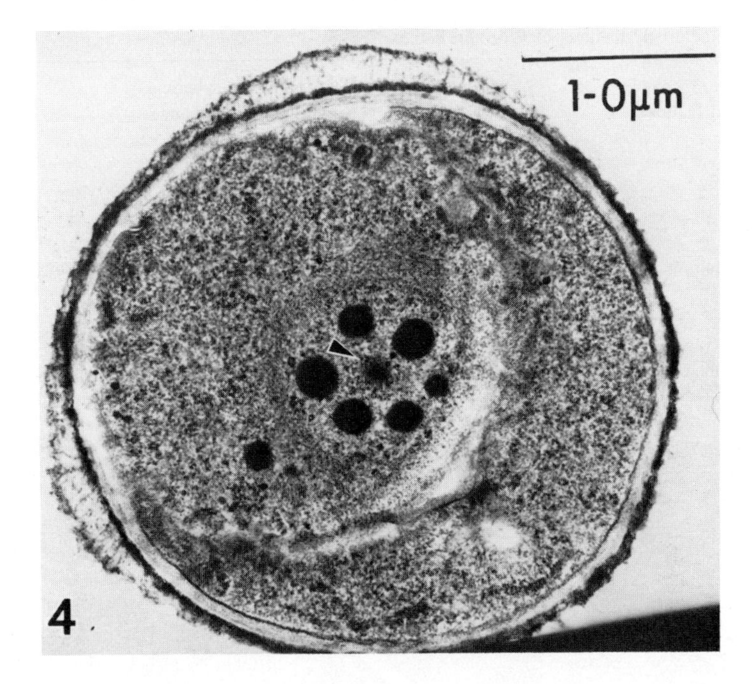

Figure 4. Transmission electron micrograph of a transverse section through an *Aspergillus giganteus* hypha, very close to a septum. The septal pore (arrowed) can be seen at the centre of the section, around which a ring of six Woronin bodies is arranged. There is a seventh Woronin body positioned alone, some distance from the central ring. This micrograph must be viewed with some caution as it was obtained from a culture grown on medium containing 0.05% w/v of the microtubule inhibitor Benlate. Nevertheless, it is remarkably similar to the micrograph published by Scannerini (1968) showing a similar arrangement of Woronin bodies in *Tuber magnatum*.

Table 1. Diameters of Woronin bodies and septal pores in *Penicillium chrysogenum* and *Aspergillus giganteus*.

Species	*Woronin body diameters (nm)*		*Septal pore diameters (nm)*	
	Mean	*Maximum*	*Mean*	*Maximum*
Penicillium chrysogenum	167 ± 8	210	88 ± 3	145
Aspergillus giganteus	216 ± 10	264	114 ± 14	165

Figures for each species were determined from a total of 100 measurements. The error limits are one standard deviation. Data from Collinge (1986).

have a consistent size in relation to pore diameter. Typical diameter measurements are given in *Table 1*. Such figures show that whilst the range of diameters of Woronin bodies from a wide variety of species is large, the variability between organelles in the same species is extremely low. Additionally, in any given species the mean Woronin body diameter is always significantly greater than the maximum septal pore diameter. The quoted data will underestimate both of these features because the measurements used to calculate the mean Woronin body diameters will inevitably include a considerable number of sub-median sections with the consequence that the true mean diameter will

Table 2. Characteristic features of Woronin bodies.

Shape	Approximately spherical
Ultrastructural features	Single membrane-bound Electron-dense
Size range	150–750 nm in diameter
Composition	Probably proteinaceous
Density	1.214 $g.cm^{-3}$
Position	Close to septal pore
Species distribution	Filamentous asomycetes and deuteromycetes only

All characteristics are quoted for the full range of species in which Woronin bodies have been observed and are either the range of possible values for a numerical feature or the most generally observed characteristic. In the case of density a figure is only available for a single species, *Penicillium chrysogenum*.

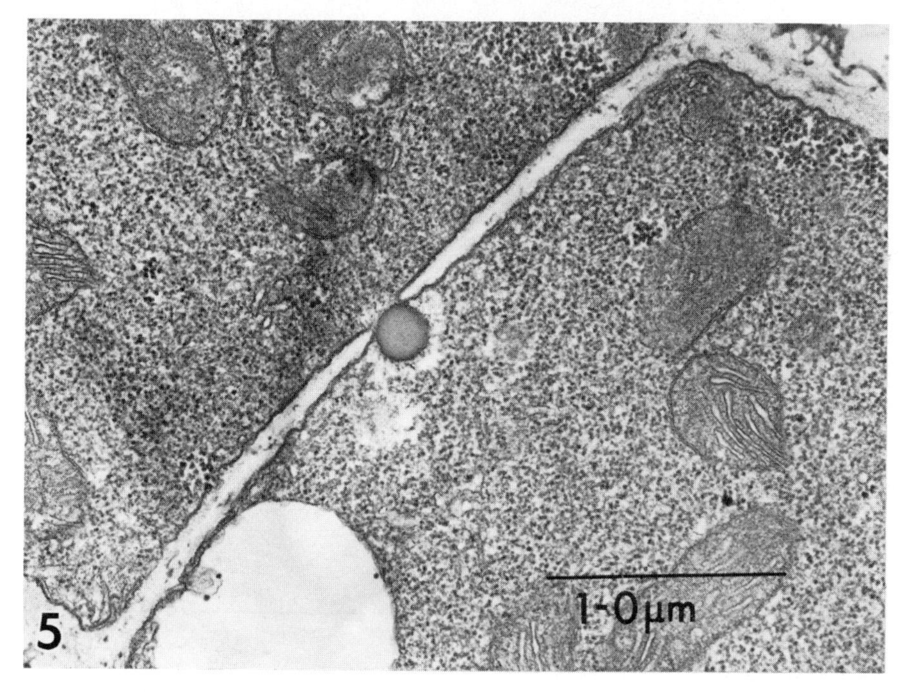

Figure 5. Transmission electron micrograph of an *Aspergillus giganteus* hypha, showing a septum in which the central pore is plugged by a Woronin body. The septum separates two apparently normal compartments.

be greater than that calculated and the true variability in diameter smaller.

A summary of the major characteristics of Woronin bodies is given in *Table 2*.

Possible roles for Woronin bodies in vegetative hyphae

The position of Woronin bodies in hyphae and their size in relation to the diameter of the septal pore suggests possible roles for these organelles in the spatial organization of hyphae involving the occlusion of the septal pore by a Woronin body. This type of thinking has been further reinforced by the frequent observation of the particular positioning of Woronin bodies illustrated in *Figure 5*. It is clear from this sort of observation that Woronin bodies do, under some circumstances, plug septal pores. This led to two important suggestions of possible roles for Woronin bodies in vegetative hyphae.

(i) In response to severe hyphal damage, they plug septal pores to prevent excessive loss of cytoplasm and therefore maintain spatial organization of hyphae.

(ii) They regulate the spatial organization of actively growing mycelia by controlling the distribution of cytoplasm and organelles within hyphae, moving to block and unblock septal pores to do so.

Maintenance of spatial organization in damaged hyphae

The idea that Woronin bodies function as emergency pore plugs in response to hyphal damage and potential spatial disruption by loss of cytoplasm has been firmly held for some time (Reichle and Alexander, 1965), but until recently it has been based on largely circumstantial evidence. There are at least two ways of testing this hypothesis, direct observation of living material by light microscopy, or the examination of fixed material by transmission electron microscopy. The first option would appear to be preferable and has offered some valuable support for the hypothesis as shown in *Figure 6*. Observation of hyphae cut with a razor blade shows quite clearly that the extent of cytoplasmic loss resultant from such severe damage is extremely limited. It is also clear that the site of limitation is the septum. This is apparent either by prior observation and recording of septal positions, or from the fact that complete collapse of a hyphal tip is apparent if the cut is made beyond the most apical septum, but cytoplasmic loss is limited very soon behind the cut by the first septum. This type of observation has provided good supporting evidence for the hypothesis, because it is clear that some mechanism to prevent cytoplasmic loss operates very quickly at the septum and logically this must be as a result of some kind of plugging of the septal pore. What is not obvious from this sort of experiment is the identity of the plug. Under such conditions particles can certainly be observed to move to the septum from nearby, and it has been claimed that these can be identified as Woronin bodies (Brenner and Carroll, 1968). However, in view of the size of these organelles the plugs could equally well be other similar sized organelles or inclusions.

The second line of evidence is from transmission electron microscopy and observations such as that shown in *Figure 7* support the hypothesis. This type of observation has been made many times and shows quite clearly that Woronin bodies do indeed plug septal pores in septa, dividing reasonably healthy compartments from ones that have been subjected to some considerable degree of loss of cytoplasm.

So for many years the suggestive evidence was good but not conclusive, because it did not link the positive identification of Woronin bodies plugging septal pores as identified by electron microscopy to a study of the response of a fungal culture to specifically induced damage at an identifiable site. Such essential evidence was provided by two separate studies carried out in our own laboratory using an essentially similar experimental procedure. The first study was carried out by Trinci and Collinge (1974) using *N. crassa*, which contains hexagonal crystals. The most recent study examined two species containing spherical Woronin bodies (Collinge and Markham, 1985; Collinge, 1986). The experimental system involved the growth of fungal colonies of

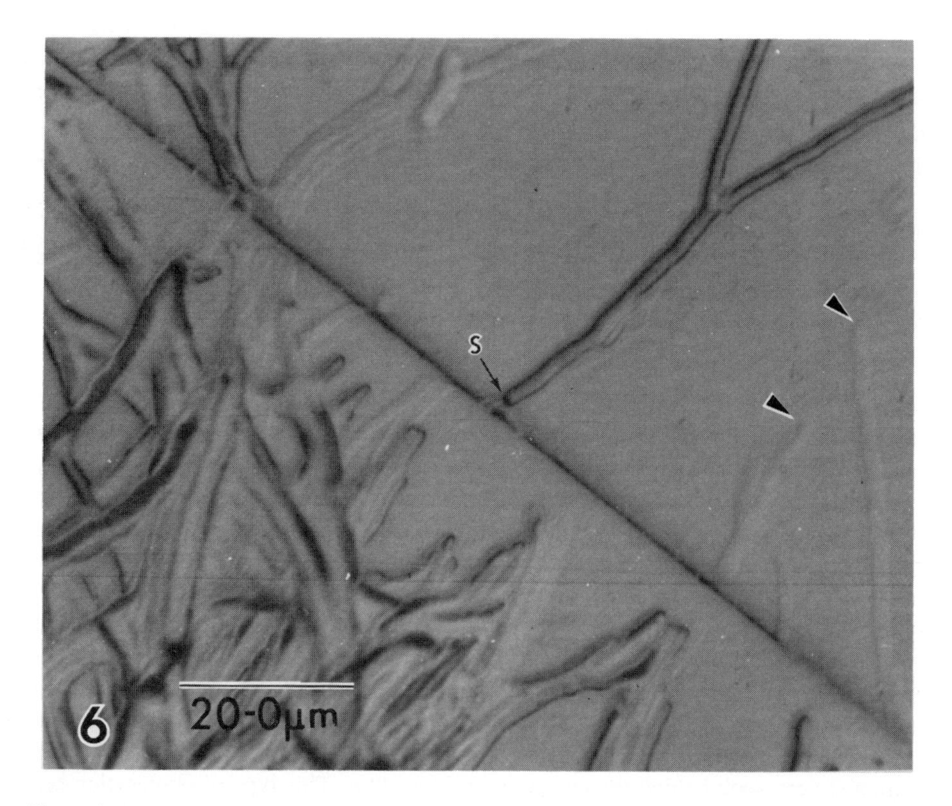

Figure 6. Hyphae at the edge of a colony of *Penicillium chrysogenum* which had been severed with a razor blade approximately 3 min before this light micrograph was taken under Nomarski differential interference contrast optics. The line of the razor blade cut can be seen running diagonally across the field of view. Several 'ghosts' of hyphal tips can be seen (arrowed) indicating tip compartments from which cytoplasm has been completely lost as a consequence of their having been severed nearer to the hyphal apex than the position of the first septum. One hypha severed further back from the tip shows retention of cytoplasm except in the immediate vicinity of the cut. The cytoplasm is retained by a septum(s) just beyond the line of the razor blade cut.

convenient size on solid medium. The simple expedient of severing the colonies with a razor blade was used to inflict damage. The cut was made very close to the edge of the colony in the region of young actively growing hyphae. The site of damage could be located when the specimen was processed for examination under the transmission electron microscope. The region of the colony in the immediate vicinity of the damage (up to 1.5 mm on either side of the cut) was examined for evidence of pore plugging. The type of observation that was made is illustrated in *Figure 8* which clearly shows that Woronin bodies do plug septal pores in response to such damage.

Interestingly, basidiomycetes achieve the same result by a different but apparently analogous mechanism of rapid deposition of electron-dense material into the septal pore channel, as has been demonstrated in *Coriolus versicolor* and *Schizophyllum commune* (Aylmore *et al.*, 1984), and *Lepiota procera* (O.Steele-Mortimer and P.Markham, unpublished results), as shown in *Figure 9*.

To test fully the hypothesis of Woronin body function, quantitative data had also

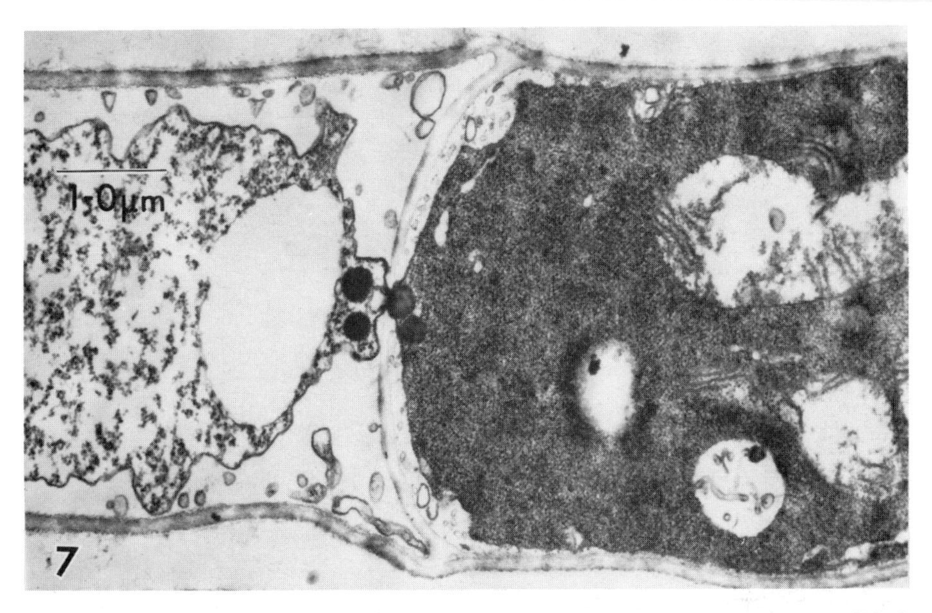

Figure 7. Transmission electron micrograph of a *Penicillium chrysogenum* hypha in which a Woronin body is shown plugging the central pore of a septum separating a compartment emptied of most of its cytoplasm from a compartment in which the bulk of cytoplasm has been retained.

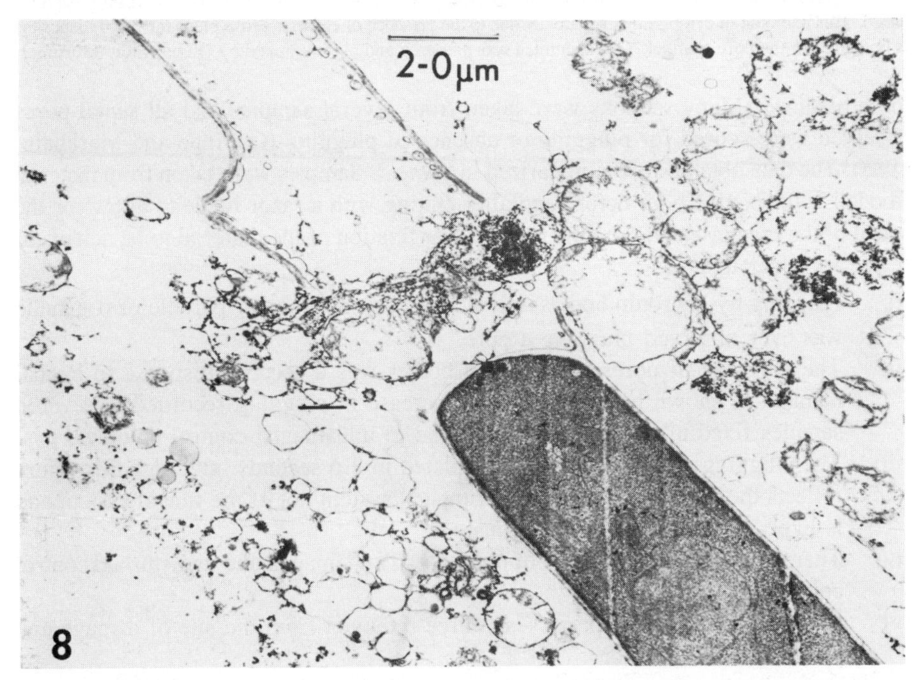

Figure 8. Transmission electron micrograph of a *Penicillium chrysogenum* hypha which had been severed with a razor blade ten seconds prior to fixation. The line of the cut can be seen crossing one compartment, from which virtually all of the cytoplasm has been lost. A Woronin body is shown plugging the pore of the septum which delimits the adjacent compartment, within which the cytoplasm has been retained and is of completely normal appearance.

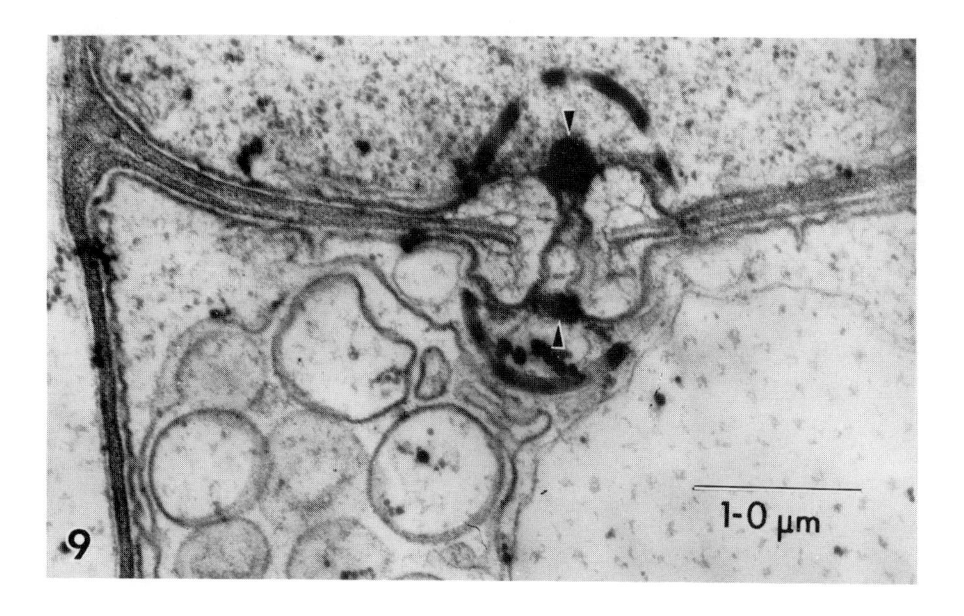

Figure 9. Transmission electron micrograph of a *Lepiota procera* hypha which had been severed with a razor blade 30 s prior to being fixed for examination. The dolipore septum is shown separating a compartment emptied of most of its cytoplasm from the adjacent compartment in which the cytoplasm has been retained. This retention of cytoplasm is apparently due to the presence of electron-dense plugs (arrowed) deposited at both ends of the pore channel. This specimen was prepared and photographed by Olivia Steele-Mortimer.

to be obtained. Many sections were taken from several samples and all septal pores observed were scored for plugging or absence of plugging (Collinge and Markham, 1985). The data obtained are summarized in *Table 3*. Samples were taken from material flooded with fixative both before and after cutting with a razor blade to allow for the speed of the response and the time required for fixation of the material to be achieved. It became clear that:

(i) Plugging by Woronin bodies only was observed; no other particle or organelle was ever observed plugging a pore.

(ii) The plugging of pores by Woronin bodies was a specific response to hyphal damage as shown by the enormous increase in plugging recorded in severed samples fixed after cutting as compared to undamaged control colonies.

(iii) The plugging process could be completed in 3.6 seconds, although this figure reflects the limitations of the experimental system not of the fungal system and is likely to be a severe overestimate.

(iv) There was a small basal level of plugging in undamaged and undisturbed control colonies.

(v) Plugging extends to septa at least three removed from the site of damage and probably much further.

It was also necessary to determine what, if anything, followed this initial rapid plugging response. This was examined by Collinge and Markham (1987) who severed colonies, but incubated them for much longer periods, 3 or 17 h, before fixing them for electron

Table 3. Percentage plugging of septal pores by Woronin bodies in colonies of *Penicillium chrysogenum* which were undamaged or cut with a razor blade.

Time from cutting to flooding with fixative (s)	*Number of septa*		*Percentage plugging*
	Plugged	*Unplugged*	
Undamaged	4	71	5.3
10.0	60	0	100.0
1.8	60	4	93.8
−1.8	32	26	55.2
−10.0	4	63	6.0

Negative values for the time from cutting to flooding with fixative indicate that the fixative was poured onto the colony before cutting was carried out. The time period during which the plugging process can be completed is from the cutting of hyphae to the addition of fixative, plus the time required for fixation to occur, as determined from samples flooded with fixative prior to cutting of hyphae (provided there is a difference in the percentage plugging in the relevant samples). On this basis the data show that plugging can be completed within 3.6 s, which is the 1.8 s elapsed before fixative was added to a cut sample plus a further 1.8 s for fixation as demonstrated by the sample flooded with fixative 1.8 s prior to having been cut. Data from Collinge and Markham (1985).

microscopy. After 3 h of further incubation the plugged septa all looked as shown in *Figure 10*. This shows that there is a slower response mounted after the initial rapid plugging and, in a process termed consolidation, material of a very similar appearance to that making up the hyphal wall and septal plate is deposited on the side of the septum where the Woronin body is positioned, that is, the side furthest from the cut region. Such deposition was never observed on the other side of the septum. This presumably represents the deposition of polymeric material over the Woronin body and septum to strengthen and make permanent the plug so that the septum remains fully sealed. Similar structures have been reported in apparently undamaged cultures of *Drechslera sorokiniana* (Cole, 1973, 1986) and an apparently equivalent process involving hexagonal crystals has been recorded in *N. crassa* (Trinci and Collinge, 1974).

Regulation of spatial organization in actively growing hyphae

It is clear then that Woronin bodies plug septal pores in response to severe damage of hyphae. What is not yet known is how severe that damage has to be to precipitate the response and this question is important in the context of the second hypothesis for the function of Woronin bodies, namely that they may regulate cytoplasmic flow and organelle distribution in hyphae by selectively plugging and unplugging septal pores. That open septal pores allow complete migratory freedom to cytoplasm and all organelles in hyphae has been confirmed by light (Buller, 1933) and electron microscopy (Shatkin and Tatum, 1959; Hunsley and Gooday, 1974). As has been observed many times, even the largest organelles in hyphae, the nuclei, can pass freely through the relatively narrow septal pore in ascomycete and deuteromycete hyphae (*Figure 11*). Septa therefore clearly do allow continuity between the contents of adjacent compartments and it is generally considered that this is due to the necessity for a co-ordination of metabolic and physiological activity between several such compartments in the most metabolically active regions of the mycelium, near the growing edge of the colony. Thus Trinci (1971)

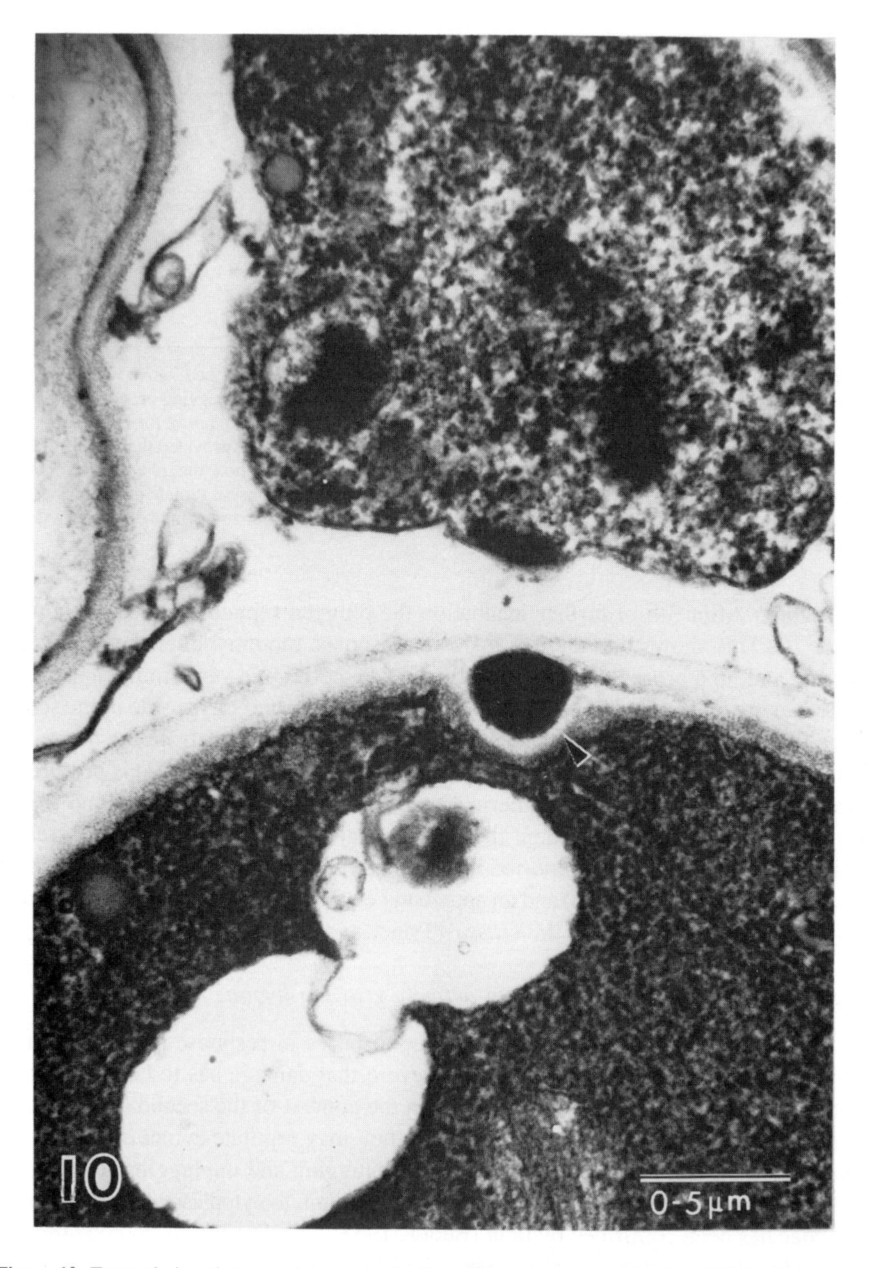

Figure 10. Transmission electron micrograph of a *Penicillium chrysogenum* hypha which had been severed with a razor blade then incubated for three hours prior to fixation for examination. The septal pore is plugged by a Woronin body and this plug has been consolidated by the deposition over the Woronin body and surrounding septum of electron-transparent material (arrowed) similar in appearance to that of which the septal plate is constructed.

has demonstrated that the supply of vesicles for the extending hyphal tip must be produced by a length of hypha identified as the peripheral growth zone, which is considerably longer than the apical compartment, itself being the longest compartment in the

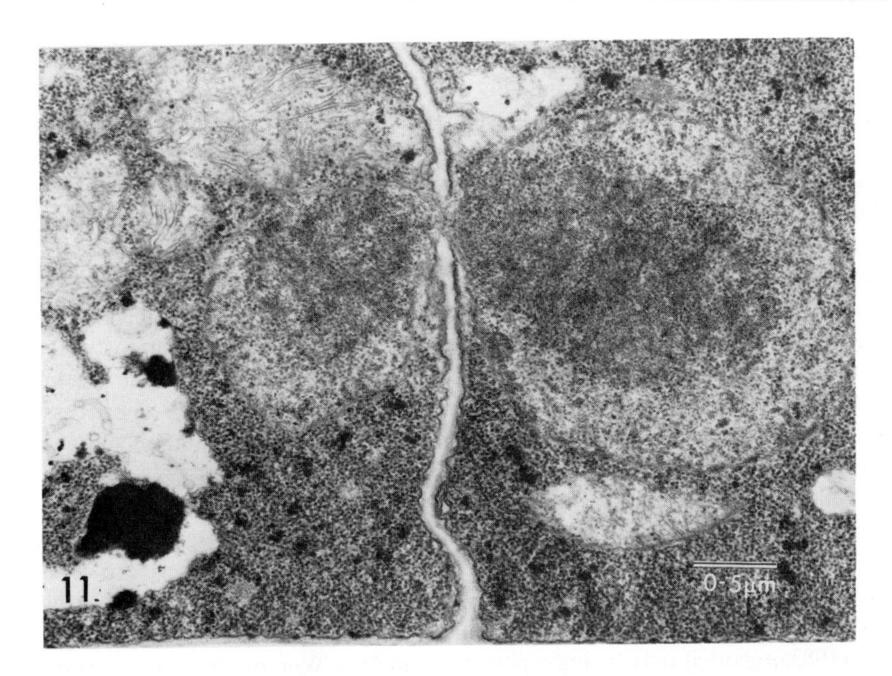

Figure 11. Transmission electron micrograph of a *Sordaria brevicollis* hypha showing a nucleus passing through the central pore of a septum.

mycelium. For example, Trinci reported that in *P. chrysogenum* strain WIS 54-1255 grown at 25°C on a defined medium, the average apical and intercalary compartment lengths were 150 μm and 29 μm, respectively. However, the peripheral growth zone extended for 500 μm and therefore corresponded to a stretch of hypha which would include more than thirteen compartments. Continuity between compartments is clearly therefore of great importance to effective metabolic and physiological activity of the mycelium. This does provide a logical basis for the hypothesis that Woronin bodies might regulate this metabolic and physiological activity by selectively opening and closing pores. However, there is as yet very little good evidence in support of the hypothesis and nothing that could not equally plausibly be explained in other ways.

Many authors have made suggestions along the same lines, but the most influential piece of work in sustaining this hypothesis for a number of years was that reported by McKeen (1971) on Woronin bodies in *Erysiphe graminis*. The most important point that he made in connection with the hypothesis under discussion is the observation that 80% of septal pores in a vigorously growing colony were plugged by Woronin bodies. On the face of it this seems quite compelling evidence for a role for Woronin bodies in plugging septal pores in undamaged, actively growing mycelia. However, examination of what McKeen says and particularly the electron micrographs he has published bring his whole hypothesis into question. Of critical importance is the fact that his micrographs show that most of the plugged septa he observed separate an apparently normal compartment from one that is much less densely packed with cytoplasm. In fact McKeen reports that 'When Woronin bodies were in septal pores the adjacent cells had a different density and structure'. Now it is conceivable that this difference in density between the divided compartments resulted from the regulatory effect of the Woronin body plug. However, this seems extremely unlikely and a much more plausible explanation is that the plugging observed by McKeen resulted from some kind of unrecorded shock to which the fungus had been subjected, causing the emergency Woronin body plugging of septal pores to be activated. This seems especially plausible because the material he used was mycelium of a plant pathogen growing on its plant host. So for example, the removal of fungal material from the plant would necessarily lead to breaks between haustoria and mycelium, which is known to cause septal plugging (Gil and Gay, 1977). If such detachment occurred before adequate fixation was completed, the emergency plugging response would be activated; indeed such specimens are notoriously slow to fix. Additionally, because it is not known how severe damage needs to be to precipitate the plugging response, it may be that breaching of the hyphal wall is not necessary. Much milder shocks such as changes in osmotic pressure or temperature shifts might activate the emergency plugging response.

Another striking point is the fact that this report by McKeen is the only one which records such a high level of plugging in actively growing 'undamaged' mycelia. One problem is that there is very little quantitative data available but there are certainly several studies which have recorded relatively low levels of plugging in actively growing mycelia. A summary of much of the available data is given in *Table 4*. Martin and Gay (1983) reported only a single pore plugged by a Woronin body in an extensive study of *Erysiphe pisi*. They did not report total numbers of septal pores observed, but in their published micrographs there are seven further unplugged pores, demonstrating a much lower percentage plugging than that observed by McKeen. This further suggests that his high figure is misleading, not even being explicable as a phenomenon characteristic of the genus, in view of the difference in this closely related species. The observations of Martin and Gay (1983) are, however, in line with all other observations of Woronin body plugging of septal pores in actively growing hyphae and with quantitative data obtained in *P. chrysogenum* (Collinge and Markham, 1985) and *Aspergillus giganteus* (Collinge, 1986). Experiments examining the response of colonies of these species to damage inflicted with a razor blade also incorporated control colonies, which were unsevered, but in every other respect treated the same as the severed colonies. Thus, quantitative estimates of the percentage plugging by Woronin bodies of septal pores in unshocked and undamaged hyphae were obtained. Particular care had been taken to ensure that these colonies were entirely undisturbed. Again the data (see *Table 4*) show that Woronin body plugging of septal pores does occur in actively

Table 4. Plugging of septal pores by Woronin bodies in undamaged actively growing mycelia.

Organism	*Number of septal pores*		*% plugging*	*Authority*
	Plugged	*Unplugged*		
Erysiphe graminis	?	?	80	McKeen (1971)
Erysiphe pisi	1	≥ 7	≤ 12.5	Martin and Gay (1983)
Penicillium chrysogenum	4	71	5.3	Collinge and Markham (1985)
Aspergillus giganteus	14	97	12.6	Collinge (1986)

? indicates figure not quoted by original author.

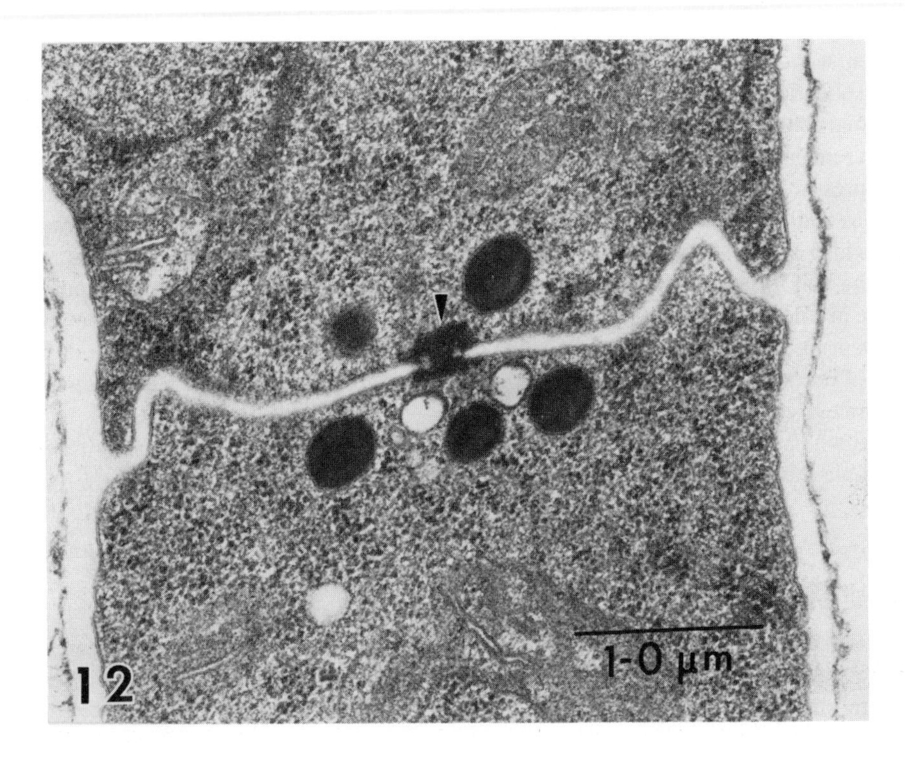

Figure 12. Transmission electron micrograph of a *Penicillium chrysogenum* hypha showing a septum in which the central pore is closed by a deposition plug (arrowed). Note the difference in appearance from that of the Woronin bodies which are positioned around the septum, particularly the bounding membrane of the Woronin bodies which is absent from the deposition plug.

growing regions of colonies, but at a much lower level than that reported by McKeen. The cause or function of this plugging is not clear. It may represent a highly selective plugging of septal pores for regulatory purposes, but equally, it could be caused by an emergency plugging response mounted as a result of localized disturbances in the mycelium, such as tip bursting, localized build-up of metabolic products causing osmotic changes and similar eventualities. It is true that our own data were obtained exclusively from the young actively growing region of the colony and older regions of the colony are far more extensively plugged, but with a different type of plug (*Figure 12*). This has been termed a deposition plug (Collinge, 1986) and is clearly distinct from a Woronin body, being far less regular in shape and not membrane-bound. They appear to result from progressive deposition of electron-dense material in the pore. Wergin (1973) has suggested that this material may be produced by Woronin bodies but there is no evidence to support this hypothesis and it seems just as likely that these plugs result from a quite separate process in which electron-dense material is deposited in and around the septal pore.

Crucially there is also a lack of evidence demonstrating a critical property that Woronin bodies would need to exhibit to carry out a regulatory function, the ability to move not only into but also out of a septal pore. Indeed the ability to move out would seem to be contrary to their apparently major function of plugging pores in response to damage, when it would be crucial that they did not become dislodged. We have observed consolidation of plugs in severed hyphae even when those plugs separate apparently normal compartments. Such consolidation has not been observed in undamaged hyphae, so it remains possible that in damaged hyphae, Woronin bodies could be held in position by pressure differences that would not occur in normal actively growing hyphae.

The ability to move in and out of septal pores would be dependent upon the mechanism by which Woronin bodies are moved within hyphae. The simplest possibility, as a result of the bulk flow of cytoplasm, would seem to exclude the option of moving out of a pore in a normal growing hypha. The alternative hypothesis of some kind of contractile system, possibly microtubules, would offer the possibility of movement both into and out of pores. However, there is no clear indication which, if either, of these mechanisms operates.

So whilst the possibility of such a regulatory function of Woronin bodies remains feasible there is really no convincing evidence to support the hypothesis.

Possible roles for Woronin bodies in sporogenesis

Though it appears that Woronin bodies do not have a regulatory role in undamaged vegetative hyphae, it is clear that in some species at least, they are important in asexual spore production.

Pore plugging at spore release

Many of the reports of Woronin bodies in different species come from ultrastructural analyses of conidial production. Consequently there is a superficial impression that Woronin bodies are generally important in the process of asexual spore production. Many hypotheses have been advanced for their role as regulators of cytoplasmic flow and organelle migration through conidiogenous cells and into developing conidia as well as for the more obvious function of plugging septal pores when conidia separate from their mother cells (Cole and Samson, 1979). However, when a careful examination of the literature is made it becomes clear that the number of species in which Woronin bodies are important in this context is relatively small; certainly there seem to be a majority of species in which they are not important. For example, species that produce conidia by phialidic development (which means synthesized entirely *de novo* by extrusion from a specialized cell called a phialide, generally in a continuous chain) very rarely have Woronin bodies in the phialide let alone the spores (Cole and Sampson, 1979). One exception is *Ceratocystis adiposa* (Hawes and Beckett, 1977). This type of conidial production system is one of the most commonly encountered, being employed by major genera such as the *Aspergilli* and *Penicillia*, in which Woronin bodies are numerous and important in the vegetative hyphae. There are, however, some species in which Woronin bodies are important in the separation of conidia from the conidiogenous cell, for example *Cercospora beticola* (Pons *et al.*, 1985). This species illustrates the importance of Woronin bodies in the separation of conidia from the conidiogenous cell in some cases. It can be clearly seen from the electron micrographs published by Pons *et al.* (1985) that in this species Woronin bodies plug the septal pore both at the base of the conidium and in the top of the conidiogenous cell. In this organism, however, each conidium is produced separately from a different site on the conidiogenous cell and becomes detached singly. This then may be the key to the importance of Woronin bodies here as compared to their non-involvement in phialidic and similar types of spore production where the mature conidia are held together in chains. In the case of *Cercospora beticola* and other species that employ Woronin bodies to seal the septal pore at the base of seceding conidia there is essentially a catastrophic separation of one cell or compartment from another, which seems to be entirely analogous to the events that occur at the septa of a hypha severed by a razor blade. This is in contrast to conidia produced in continuous chains, which often are not subjected to separation until some considerable time after they have been synthesized and therefore are not subjected to a catastrophic event. This suggestion is further supported by the fact that in *Scopulariopsis brevicaulis*, a species in which Woronin bodies do seal the basal septa of conidia, such plugged pores have been found to be consolidated by the deposition of material around the organelle just as has been reported for damaged vegetative hyphae, although in this instance, the deposited material was electron-dense (Cole and Aldrich, 1970; Cole, 1986).

Regulation of the flow of cytoplasm and migration of organelles through conidiogenous cells and into spores

A role for Woronin bodies in the plugging and unplugging of the basal septa of phialides and other conidiogenous cells, thus regulating organelle and cytoplasmic movement, has been implied (Cole and Samson, 1979). However, the evidence for this suggestion is as poor as that on which the same hypothesis for vegetative hyphae is based. There have been a few reports of such plugging of basal septa, but there are far more reports of open septa in extensive studies of conidiogenous cells. Even in cited reports of plugging (Cole and Samson, 1979), examination of the original electron micrographs generally shows close association of Woronin bodies with the pore but not plugging, or that the plug is not a Woronin body, but something similar to a deposition plug or even more specialized structures [for example at the phialide base in *Trichoderma saturnisporum* (Hammill, 1974)]. There is some evidence for a regulating pore plug at the base of ascal and ascogenous cells, for example in species of *Ascobolus*, *Saccobolus* (Carroll, 1967) and *Sordaria fimicola* (Furtado, 1971), but here the evidence is good because of a consistent observation of highly structured and unusual pore plugs that appear to form some kind of sieve-element over the septal pore. The evidence therefore suggests that when there is some requirement for a regulatory plug, a specialized structure is formed and such regulation is not performed by the preformed Woronin body plugs.

It is our view that the role of Woronin bodies in conidiogenesis and conidial secession, where they are important, is not significantly different from their role in vegetative hyphae. They are simply responsible for plugging septal pores in response to some catastrophic disturbances of the cytoplasmic continuum. Further, the fact that many conidial production systems do not involve Woronin bodies at all strengthens the case against the hypothesis that Woronin bodies have a regulatory role in spatial distribution of cytoplasm and organelles, because where more than in the production of separable viable units should an organelle responsible for such regulation be involved?

Conclusions

In answer to the question addressed by this chapter, Woronin bodies do function to maintain the spatial organization of fungal hyphae, but only when the integrity of the cytoplasmic continuum is threatened by severe stress. There is, however, no evidence to support the hypothesis that Woronin bodies exert any regulatory control over the spatial distribution of cytoplasm or organelles within actively growing hyphae. This remains a feasible but unlikely hypothesis, and all relevant observations can be readily explained in terms of a single role for Woronin bodies in an emergency septal pore plugging response to catastrophic damage or disruption of hyphae. This would include both unpredictable environmental damage sustained by the mycelium and the catastrophic events that occur at natural separation of some conidiospores from the cells which produce them.

Acknowledgements

We are grateful to Dr J.L.Gay for useful discussion of some of the points considered here. We thank the SERC for grant GR/C/33796 and a studentship to Josephine B.Head.

References

Allen,E.D. (1976) Development of crystalline inclusions ('ergosterol crystals') in *Neurospora crassa. Protoplasma,* **90**, 297–306.

Avers,C.J. (1971) Peroxisomes of yeast and other fungi. *Sub-Cellular Biochemistry,* **1**, 25–37.

Aylmore,R.C., Wakley,G.E. and Todd,N.K. (1984) Septal sealing in the basidiomycete *Coriolus versicolor. Journal of General Microbiology,* **130**, 2975–2982.

Brenner,D.M. and Carroll,G.C. (1968) Fine-structural correlates of growth in hyphae of *Ascodesmis sphaerospora. Journal of Bacteriology,* **95**, 658–671.

Buller,A.H.R. (1933) *Researches in Fungi. Volume V.* Hafner, New York.

Carroll,G.C. (1967) The fine structure of the ascus septum in *Ascodemis sphaerospora* and *Saccobolus kerverni. Mycologia,* **59**, 527–532.

Cole,G.T. (1973) Ultrastructure of conidiogenesis in *Drechslera sorokiniana. Canadian Journal of Botany,* **51**, 629–638.

Cole,G.T. (1986) Models of cell differentiation in conidial fungi. *Microbiological Reviews,* **50**, 95–132.

Cole,G.T. and Aldrich,H.C. (1970) Ultrastructure of conidiogenesis in *Scopulariopsis brevicaulis. Canadian Journal of Botany,* **49**, 745–755.

Cole,G.T. and Samson,R.A. (1979) *Patterns of Development in Conidial Fungi.* Pitman, London.

Collinge,A.J. (1986) Emergency response of ascomycete and deuteromycete fungi to specific hyphal damage. Ph.D. Thesis, University of London.

Collinge,A.J. and Markham,P. (1985) Woronin bodies rapidly plug septal pores of severed *Penicillium chrysogenum* hyphae. *Experimental Mycology,* **9**, 80–85.

Collinge,A.J. and Markham,P. (1987) Response of severed *Penicillium chrysogenum* hyphae following rapid Woronin body plugging of septal pores. *FEMS Microbiology Letters,* **40**, 165–168.

de Duve,C. (1971) Tissue fractionation past and present. *Journal of Cell Biology,* **50**, 200–550.

Furtado,J.S. (1971) The septal pore and other ultrastructural features of the pyrenomycete *Sordaria fimicola. Mycologia,* **63**, 104–113.

Gil,F. and Gay,J.L. (1977) Ultrastructural and physiological properties of the host interfacial components of haustoria of *Erysiphe pisi in vivo* and *in vitro. Physiological Plant Patholology,* **10**, 1–12.

Gull,K. (1978) Form and function of septa in filamentous fungi. In *The Filamentous Fungi.* Smith,J.E. and Berry,D.R. (eds), Academic Press, New York, Vol. 3, pp. 487–573.

Hammill,T.M. (1974) Septal pore structure in *Trichoderma saturnisporum. American Journal of Botany,* **61**, 767–771.

Hawes,C.R. and Beckett,A. (1977) Conidium ontogeny in the *Chalara* state of *Ceratocystis adiposa.* II. Electron microscopy. *Transactions of the British Mycological Society,* **68**, 267–276.

Hawksworth,D.L., Sutton,R.C. and Ainsworth,G.C. (1983) *Ainsworth and Bisby's Dictionary of the Fungi.* 7th edition. Commonwealth Mycological Institute, Kew, UK

Heath,I.B. (1977) Morphology and ontogeny of microbodies in the oomycete fungus *Saprolegnia elongatus. Canadian Journal of Botany,* **55**, 556–562.

Hoch,H.C. and Maxwell,D.P. (1974) Proteinaceous hexagonal inclusions in hyphae of *Whetzelinia sclerotiorum* and *Neurospora crassa. Canadian Journal of Microbiology,* **20**, 1029–1035.

Hunsley,D. and Gooday,G.W. (1974) The structure and development of septa in *Neurospora crassa. Protoplasma,* **82**, 125–146.

Kobr,M.J., Vanderhaeghe,F. and Combepine,G. (1969) Particulate enzymes of the glyoxylate cycle in *Neurospora crassa. Biochemical and Biophysical Research Communications,* **37**, 640–645.

Lloyd,D. (1974) Subcellular fractionation of eukaryotic microorganisms: the isolation of mitochondria and their subfractionation. In *The Mitochondria of Microorganisms.* Lloyd,D. (ed.), Academic Press, London, pp. 54–81.

Lloyd,D. and Cartledge,T.G. (1974) Marker enzymes of eukaryotic microorganisms. In *Methodological Developments in Biochemistry.* Reid,E. (ed.), Longman, London, Vol. 4, pp. 339–346.

Markham,P. and Collinge,A.J. (1987) Woronin bodies of filamentous fungi. *FEMS Microbiology Reviews,* **46**, 1–11.

Martin,M. and Gay,J.L. (1983) Ultrastructure of conidium development in *Erysiphe pisi. Canadian Journal of Botany,* **61**, 2472–2495.

Mason,P.J. and Crosse,R. (1975) Crystalline inclusions in hyphae of the *glaucus* group of Aspergilli. *Transactions of the British Mycological Society,* **65**, 129–134.

Maxwell,D.P., Maxwell,M.D., Hanssler,G., Armentrout,V.N., Murray,G.M. and Hoch,H.C. (1975) Microbodies and glyoxylate-cycle enzyme activities in filamentous fungi. *Planta,* **124**, 109–123.

Mendgen,K. (1973) Microbodies (glyoxysomes) in infection structures of *Uromyces phaseoli. Protoplasma,* **78**, 477–482.

Muller,M., Hogg,J.F. and de Duve,C. (1971) Distribution of tricarboxylic acid cycle enzymes and glyoxylate cycle enzymes between mitochondria and peroxisomes in *Tetrahymena pyrifomis. Journal of Biological Chemistry,* **243**, 5385–5395.

McKeen,W.E. (1971) Woronin bodies in *Erysiphe graminis* DC. *Canadian Journal of Microbiology,* **17**, 1557–1560.

Pons,N., Sutton,B.C. and Gay,J.L. (1985) Ultrastructure of conidiogenesis in *Cercospora beticola. Transactions of the British Mycological Society,* **85**, 405–416.

Powell,M.J. (1976) Ultrastructure and isolation of glyoxysomes (microbodies) in zoospores of the fungus *Entophlyctis* sp. *Protoplasma,* **89**, 1–27.

Reichle,R.E. and Alexander,J.V. (1965) Multiperforate septations, Woronin bodies, and septal plugs in *Fusarium. Journal of Cell Biology,* **24**, 489–496.

Scannerini,S. (1968) Setti con 'corpi di Woronin' in *Tuber magnatum* pico. *Allionia,* **14**, 63–76.

Shatkin,A.J. and Tatum,E.L. (1959) Electron microscopy of *Neurospora crassa* mycelia. *Journal of Biophysical and Biochemical Cytology,* **6**, 423–426.

Tolbert,N.E. and Essner,E. (1981) Microbodies: peroxisomes and glyoxysomes. *Journal of Cell Biology,* **91**, 271–283.

Trinci,A.P.J. (1971) Influence of the width of the peripheral growth zone on the radial growth rate of fungal colonies on solid media. *Journal of General Microbiology,* **67**, 325–344.

Trinci,A.P.J. and Collinge,A.J. (1974) Occlusion of the septal pores of damaged hyphae of *Neurospora crassa* by hexagonal crystals. *Protoplasma,* **80**, 57–67.

Vigil,E.L. (1973) Structure and function of plant microbodies. *Sub-Cellular Biochemistry,* **2**, 237–285.

Wergin,W.P. (1973) Development of Woronin bodies from microbodies in *Fusarium oxysporum* f.sp. *lycopersici. Protoplasma,* **76**, 249–260.

Wilson,C.L., Stiers,D.L. and Smith,G.G. (1970) Fungal lysosomes or spherosomes. *Phytopathology,* **60**, 216–227.

Woronin,M. (1864) Zur Entwicklungsgeschichte des *Ascobolus pulcherrimus.* Cr. und einiger Pezizen. *Abhandlungen der Senckenbergischen Naturforschenden Gesellschaft,* **5**, 333–344.

CHAPTER 7

Spatial organization of microfilaments in amoeboid locomotion

C.A.KING and S.K.MACIVER

Department of Zoology and Cell Biology, University College London, Gower Street, London WC1E 6BT, UK

Introduction

Microfilaments (MF) are classically defined as 7 nm wide cytoskeletal filaments based on the protein actin (monomeric mol. wt = 42 000).

It has been recognized for many years that an equilibrium exists between the monomeric form of actin (globular actin/G-actin) and the fibrillar form of actin (F-actin).

$$\text{G-actin} \rightleftharpoons \text{F-actin}$$

Under suitable experimental conditions G-actin can be polymerized *in vitro* and the nature of the polymerization studied in detail (Pollard and Cooper, 1986). Under optimal conditions particularly in the presence of ATP and Mg^{2+} ions there exists a critical actin concentration (of about 1 μM) below which polymerization from G-actin to F-actin will not occur. However, within many cells the concentration of actin is in excess of 100 μM and it might be thought that most of the cellular actin would be present as microfilaments. This is not the case due to the presence of a variety of actin-binding proteins (Pollard and Cooper, 1986).

It can be shown that individual MF exhibit polarity using fragments of myosin molecules. These have been obtained by enzymic treatment of complete myosin molecules with proteases to yield heavy meromyosin (HMM) or the myosin subfragment 1 (S_1). Each S_1 molecule retains an adenosine triphosphatase (ATPase) site and an actin binding site from the original myosin molecules and, therefore, when MF are treated with S_1 fragments the filaments become 'decorated', that is distinctive lateral projections are formed. The presence of many S_1 fragments along a single MF gives the appearance of an arrowhead so that following decoration with myosin fragments two ends on the MF can be recognized: the barbed end and the pointed end by analogy to the structure of an arrow (Alberts *et al.*, 1983).

This chapter will be concerned with the role of microfilaments in amoeboid movement. Nevertheless it is important to stress that microfilaments are likely to play important roles in the overt motility of several protista, for example diatoms (Edgar and Zavortink, 1983), coccidian (Russell and Sinden, 1981) and gregarine protozoans (King *et al.*, 1986) and in intracellular cytoplasmic streaming particularly in the giant algal cells of *Nitella* and *Chara* (Allen and Allen, 1978; Sheetz and Spudich, 1983).

Amoeboid movement in protozoa

There is a wide variety of amoeboid cells in the protozoan kingdom carrying out crawling movement (Lee *et al.*, 1985) and thus it is possible that different strategies have been developed in various amoeboid cells. The classical description of amoeboid movement (based largely on the study of *Amoeba proteus*) identifies a clear area at the anterior end (termed the hyaline cap). It is at this point that eruption of cytoplasm usually occurs to give a pseudopodium (false foot). This newly formed pseudopodium will make contact with substrate and consolidation of this contact is clearly important in order for forward movement to occur. In the simplest case the eruption of a single bulky pseudopodium (lobopodium) leads to forward translocation by monopodial locomotion. More than one eruptive site may occur leading to multipodial locomotion. In other cases the anterior cytoplasmic extensions are provided by filopodia and lamellipodia (Taylor and Condeelis, 1979).

The giant amoebae (*Amoeba, Chaos*) have provided the experimental cells for which many models of amoeboid locomotion are based. *Amoeba proteus* is about 500 μm in length and is therefore amenable to microsurgery and microinjection. It can be shown that intracellular injection of deoxyribonuclease (DNase) I, which forms a 1:1 complex with G-actin, causes the cessation of motility (Wehland *et al.*, 1979) as does the addition of the mushroom poison phalloidin, which combines with F-actin (Stockem *et al.*, 1978). These experiments clearly show the importance of the G⇌F-actin interconversion in order for motility to occur. Nevertheless the use of the large amoebae has several disadvantages.

(i) Axenic growth of the cells has not been possible and it is therefore very difficult to get enough cells to characterize the cytoskeletal proteins present.
(ii) The presence of symbionts within the cytoplasm makes identification of cytoplasmic components more difficult.
(iii) The presence of a very extensive glycocalyx (~200 nm thick) exterior to the plasmalemma suggests that the cell-substrate interactions occurring during crawling will be very different from those occurring in most other crawling cells.

Two general models for amoeboid movement have been produced from studies on *Amoeba*.

(i) The classical work of Pantin and Mast (Mast, 1926) considered the changes in the viscous state of the cytoplasm during movement (i.e. gel–sol interconversion). It was proposed that at the advancing end of the amoeba the fluid 'sol' cytoplasm (endoplasm) is converted to a gel (ectoplasm). At the rear of the cell the peripheral ectoplasm (cortex) changed to a more fluid constitution (sol) (*Figure 1*). It was envisaged that contraction events occurred at the rear of the cell forcing the sol endoplasm forward and producing cytoplasmic streaming. This theory has sometimes been compared with squeezing a toothpaste tube. Various other theories are a modification of this general theme, for example Grebecka and Grebecki (1981) proposed that the contraction was not limited to the rear end but could take place over the whole cortical tube.
(ii) The 'frontal contraction theory' was proposed by Allen (1961). This hypothesis proposed that the endoplasmic sol was pulled forward by contractions of the

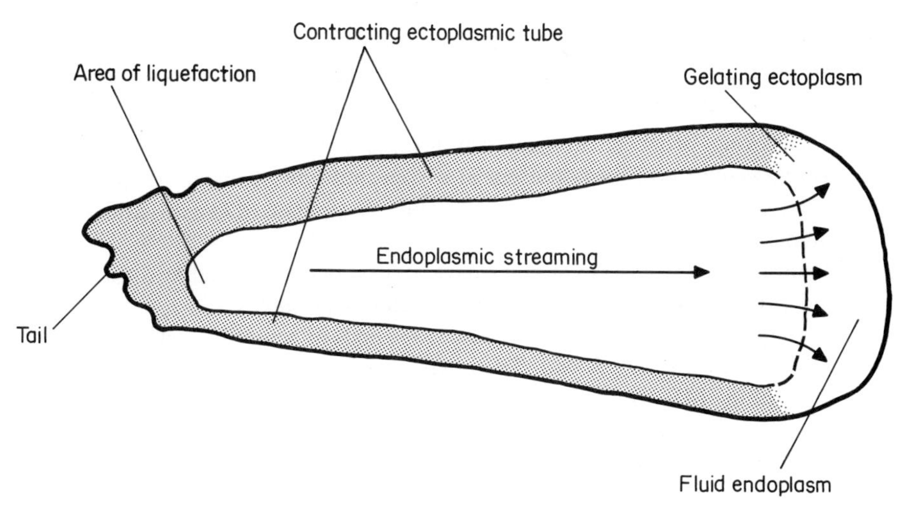

Figure 1. Sol – gel transformation in a locomoting amoeba based on the ideas of Pantin and Mast (Mast, 1926).

anterior endoplasm leading to a fountain-like flow of cytoplasm.

These two theories are clearly very different but recent work on the role and distribution of myosin in locomoting *Dictyostelium* amoebae (Yumura and Fukui, 1985) gives considerable support to the first.

There are two aspects to the problem of amoeboid locomotion, that is the cytoskeleton and the cytomusculature. With regard to the cytoskeleton it is generally accepted that the major cytoskeletal framework is the cortex (ectoplasm) and this is frequently referred to as the cortical tube. However it is also likely that cytoskeletal elements play important roles in endoplasmic regions; for example, Preston (1985) has described a microtubular 'cage' which embraces the nucleus, in *Acanthamoeba* amoebae. Linkages between the cortex and the plasmalemma will be particularly important in order that cell – substrate interactions can occur, which lead to the development of traction on the substrate – a feature of amoeboid locomotion frequently overlooked.

With respect to the cytomusculature, the interaction of thin filaments (actin microfilaments) with thick filaments (myosin filaments) in the sarcomeres of vertebrate striated muscle (Alberts *et al.*, 1983) has been considered the prime candidate for models of contraction events in amoebae. The observation that glycerol-extracted amoebae could contract when Mg^{2+}-ATP was added (Hoffman-Berling, 1956) in a similar way to that previously described for glycerol-extracted skeletal muscle, gives support for this concept. It should be noted that the amount of myosin in non-muscle cells is relatively low, for example 3.6 μM myosin (170 μM actin) in *Acanthamoeba* amoebae compared with 144 μM myosin (900 μM actin) in rabbit skeletal muscle (Pollard, 1981; Pollard and Cooper, 1986).

Central to the problem of amoeboid locomotion is the nature of the gel and sol state of the cytoplasm. In order to understand this problem we have to reconsider the G-actin/F-actin interactions. F-actin filaments alone would not produce the high viscosity cytoplasm (gel) found in the cortical ectoplasmic regions of amoebae. The formation of gels requires the presence of specific gelation factors which can combine with actin filaments to form meshworks. These factors are but one example of a group of proteins which can interact with actin and termed the actin binding proteins. Various classes of such proteins can be recognized.

(i) Sequestering proteins. Profilin can form a 1:1 complex with G-actin. In *Acanthamoeba* profilin is present at a concentration of about 100 μM and is thought to act as an 'actin-buffer' regulating the amount of G-actin available for polymerization reactions.
(ii) Bundling proteins. These proteins have at least two actin binding sites and can bind MF into linear arrays to produce bundles or cables.
(iii) Capping proteins. These proteins bind to the ends of actin filaments – generally to the barbed (preferred) end. These will clearly be important in regulating the length of the MF. The cytochalasin drugs which inhibit actin-based motility act in a rather similar way by binding to the barbed ends of MF.
(iv) Severing proteins. These proteins react with actin units in the MF bringing about breakage of actin–actin bonding and producing severing. Such proteins could clearly act as solation factors. Some of these proteins are also capping proteins.
(v) Cross linking proteins (gelation factors). These proteins can interact with MF to produce isotropic networks, the viscosity of which is similar to that found in ectoplasmic gels.

An understanding of the conditions at the front of an amoeba favouring the eruption of pseudopodia and gelation and those at the rear of the cell favouring solation, perhaps by the generation of active solation factors, would clearly be important in order to understand how actin-binding proteins play their role in polarized cell movement.

Amoeboid locomotion in *Dictyostelium discoideum*

This cellular slime mould has two forms of amoebae present in the life cycle (see Chapter 5 by Watts). Vegetative amoebae feed on bacteria and their locomotion can become highly polarized under the influence of chemotactic factors such as folic acid, presumably released by bacteria. When the amoebae are starved, they develop into aggregative amoebae. They now respond chemotactically to cyclic adenosine monophosphate (cyclic AMP) which is secreted by the amoebae themselves. This leads to the production of an aggregate of cells which will produce the slug.

The amoebae of *D. discoideum* can be grown in suspension culture and this has enabled the isolation and characterization of several cytoskeleton-associated proteins. The cortex of *D. discoideum* amoebae has been extensively studied by Condeelis (1981). It has been possible to stabilize the cortices of these amoebae by addition of the lectin concanavalin A. Following detergent treatment the fibrillar cytoskeleton of the cortex is obtained. One particular actin-binding protein (the 120 K protein) is present in the cortex at a concentration of 2.3% of the total cell protein (Condeelis *et al.*, 1984). Although this protein has a mol. wt of 120 kd when determined by SDS gel electrophoresis, it is present as a dimer (mol. wt 241 kd) when present in the cortex. In vitro, the 120 K protein plus actin filaments can form fibrillar meshworks similar to those seen *in vivo* (*Figure 2*).

As the molar ratio of 120 K protein to actin increases, the pore size (defined as the

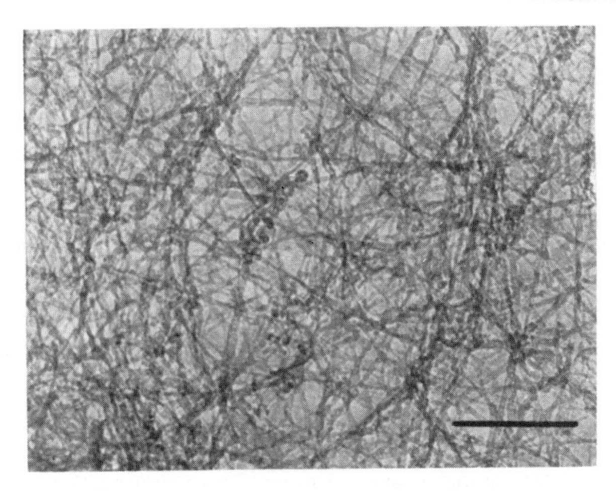

Figure 2. Electron micrograph of a critical point-dried whole mount preparation made by co-polymerizing actin and 120 K protein isolated from *D. discoideum* amoebae at a molar ratio of 14:1 followed by glutaraldehyde fixation. (Reproduced by courtesy of John Condeelis.) Bar marker = 0.2 μm.

average filament length between filament crossover points in the gel) decreases. At a molar ratio of 1:14 the pore size ranges from 24 to 56 nm and therefore a myosin monomer (150 nm in length) would diffuse only very slowly through such a lattice. Oligomers of myosin (e.g. thick filaments) could not enter such a lattice or would become trapped in the gel matrix. In either case actin–myosin interactions could not occur and therefore filament sliding (i.e. contraction) would not take place. This important concept concerning the inability of contractile events based on actomyosin to occur within the gel has been used to formulate the solation–contraction hypothesis (Hellewell and Taylor, 1979; Condeelis, 1981). Whereas actin normally stimulates the Mg^{2+}-ATPase activity of myosin, this enzymic activity is lost in the presence of the 120 K protein. The 120 K protein can also act to nucleate actin polymerization.

Another important actin-binding protein found in the cortex is a 95 K protein (native mol. wt 210 000) which has many properties in common with the α-actinin present in other cell types. It is present at a concentration of 1.9% of the total cell protein. In contrast to the 120 K protein it interacts with actin MF to produce lateral arrays (with spacing of about 40 nm between the MF) and this interaction is sensitive to the presence of Ca^{2+} ions. It is envisaged that the organization of MF into lateral arrays by 95 K protein in the cytoplasmic matrix allows efficient interaction with myosin molecules to occur, generating a force regulated by ambient Ca^{2+} (or H^+) concentrations.

Bennett and Condeelis (1984) have shown that MF present in the cortex make two types of contact with the cell membrane, viz. end-on attachment by the barbed end of the MF and lateral attachment by rod-like bridges. They calculated that about 25 000 such contacts were present in a single amoeba. It would be expected that these contact sites with the cell membrane have important roles in linking cell-substrate interactions occurring at the extracellular face of the plasmalemma with the fibrillar structure of the cortex. A considerable amount of work has been carried out on the mechanism whereby the chemotactic signal cyclic AMP produces polarized movement of amoebae (McRobbie, 1986; Fukui and Yumura, 1986). Particular importance seems to relate to the dynamics of myosin molecules. On chemotactic stimulation, myosin is translocated from the endoplasm into the ectoplasmic regions at the rear of the cell (Fukui and Yumura, 1986) where it is thought that actomyosin contraction occurs. Currently, attention is being directed to the state of myosin within this system particularly with regard to the phosphorylation of the myosin molecule and its degree of polymerization.

Amoeboid locomotion in *Naegleria gruberi*

The genus *Naegleria* comprises mostly soil-dwelling amoebae, but *N. fowleri* can be pathogenic in man causing Primary Amebic Meningoencephalitis (John, 1982) by invading the brain using its powers of amoeboid locomotion.

Our studies on *N. gruberi* amoebae have been concerned primarily with the interactions between crawling amoebae and substrate. These amoebae have seveal features that make them an excellent system for studying the problems of amoeboid locomotion.

(i) They are easily grown in culture
(ii) In contrast to the giant amoebae their size (about 20–25 μm long) is close to the average cell size displayed by eukaryotic cells.
(iii) They lack the extensive glycocalyx present on the giant amoebae.
(iv) Crawling motility can be carried out in the simplest of media – deionized water.
(v) Polarized forward movement can occur readily in the absence of chemotactic cues (in contrast to *Dictyostelium*).
(vi) Rapid rates of locomotion are observed. In 10 mM NaCl the amoebae can move at about 1 $\mu m\ s^{-1}$ over glass substrates (20°C). Doubling the speed of movement can be achieved by raising the temperature to 30°C (King *et al.*, 1983a). These rates are high compared with most other crawling cells; for example, motility rates for human polymorphonuclear leucocytes are 0.1 $\mu m\ s^{-1}$; for cultured fibroblasts 0.002–0.02 $\mu m\ s^{-1}$; for *D. discoideum* amoebae (under chemotactic stimulation) up to 0.5 $\mu m\ s^{-1}$; for *Amoeba* and *Chaos* 1–4 $\mu m\ s^{-1}$).

Development of traction on the substrate via cell-substrate interactions is a very important requirement for cell translocation. The use of reflexion interference microscopy (RIM) (Curtis, 1964) has enabled cell-substrate separation distances to be studied in living motile cells. In this technique, amoebae are allowed to attach to a glass coverslip in a suitable aqueous medium and then viewed by epi-illumination through an oil immersion objective lens. Some of the incident light will be reflected (R_1) from the glass coverslip–medium boundary due to differences in refractive index. A proportion of light waves proceeding onwards through the medium will be similarly reflected (R_2) at the interface between the medium and ventral surface of an amoeba. Using monochromatic light these two rays (R_1 and R_2) may either reinforce each other (constructive interference) or interfere with each other (destructive interference). Thus, in RIM studies if the background illumination appears grey, constructive interference would produce a white image and destructive interference would produce a dark or black image. The production of these different images will clearly be related to the

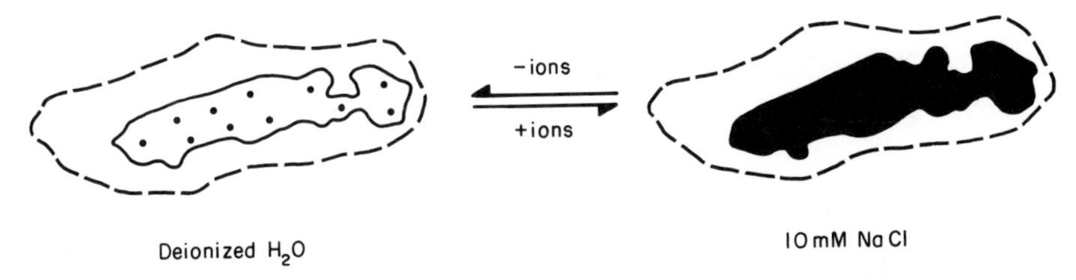

Figure 3. Diagram to show the influence of external ions on the reflexion interference microscope image obtained from the ventral surface of a *N. gruberi* amoeba moving on a glass substrate. At very low ionic strengths the associated contract (area enclosed by solid line) is held at a distance of about 120 nm from the substrate. Within this area the punctate focal contacts are shown. On raising the ionic strength of the medium the associated contact becomes darker and eventually the cell-substrate separation distance generates a black image (that is separation distance is less than 20 nm). The dotted line depicts the peripheral margin of the cell.

difference in path length between the two rays R_1 and R_2, related in turn to the cell-substrate separation distance. The technique is particularly useful for looking at cell-substrate gap distances from zero to 150 nm. In a matched RIM/transmission electron microscopic study we found good agreement using these techniques for calculation of the cell-substrate separation distances in *Naegleria gruberi* amoebae (King *et al.*, 1983b).

We have studied three interrelated aspects of traction development, described below.

(i) *Cell-substrate separation distance*

An amoeba moving on a glass substrate in 10 mM NaCl generates a black zero order reflexion interference image (i.e. gap distance less than 20 nm) from its ventral surface. This contact area (termed associated contact) is very dynamic and represents about one third of the cell profile area. On replacement of the 10 mM NaCl by deionized H_2O the image obtained from this platform area changes from black to white (the gap distance of the associated contact is now about 120 nm) (but the contact area still represents about one third of the cell profile. In deionized water, small (0.4 μm diameter) black dots can be found within the associated contact area termed 'focal contacts' (Preston and King, 1978). It is proposed that these punctate contacts are produced by eruptive filopodia produced from the ventral surface of the amoeba making contact with the substrate.

Figure 3 shows the change in contact pattern induced by changes in ionic strength of the medium. In contrast to the dynamic nature of the associated contact, the focal contacts, once formed, are fixed in position relative to the substrate. One can only show this using low ionic strength solutions because it is not possible to detect black punctate focal contacts against a black associated contact! These focal contacts are generally formed at the front of the cell and disappear at the rear of the cell (*Figure 4*). In our previous electron microscopic study (King *et al.*, 1983b) we obtained a contact area which had the properties of a focal contact i.e. it was about 0.4 μm in diameter and possessed a cell-substrate separation distance of less than 20 nm but was flanked by cell surface which was held at a distance of about 120–140 nm from the substrate (*Figure 5*).

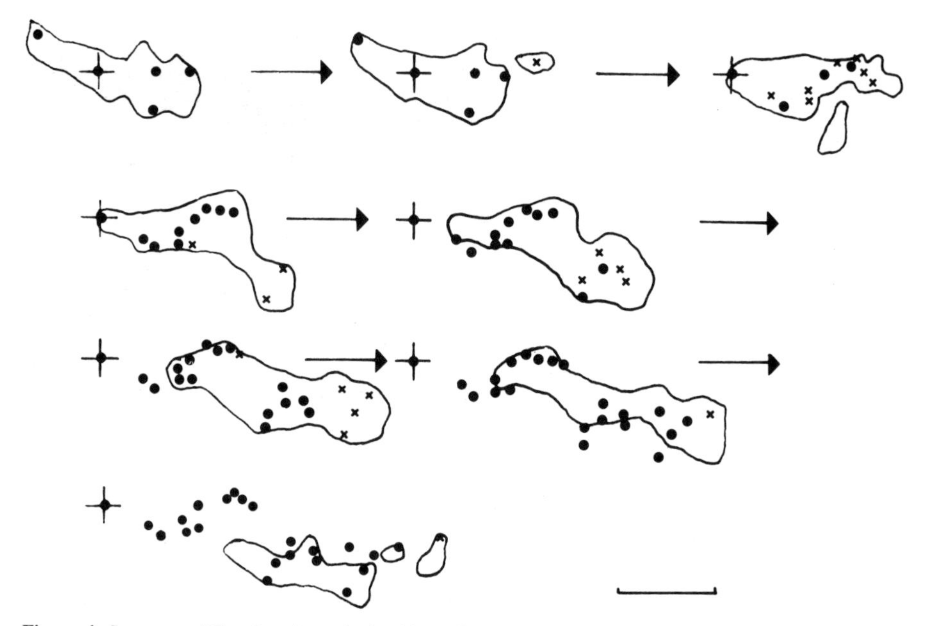

Figure 4. Sequence of line drawings obtained by reflexion interference microscopy showing the behaviour of a single *N. gruberi* amoeba moving in deionized water on a glass substrate. The top left hand diagram shows the situation at time 0 followed by a sequence of 10 s intervals. The solid line represents the boundary of the associated contact. ●, focal contact; ×, new focal contact formed in previous 10 s, ✦, one particular focal contact marked distinctively to provide a point of reference. Bar marker = 10 μm (from Preston and King, 1978).

Disappointingly, there were no cytoskeletal elements associated with the focal contact or the cortex. The fixation process was carried out while the amoeba was observed under RIM to ensure that the cell-substrate gap distance did not change during fixation. It is probable that the fixative did not enter the cell quickly enough to preserve the cytoskeletal elements. A recent paper on identification of filamentous structures in *Physarum* plasmodia stressed the importance of detergent addition in order to drive fixative into the cell sufficiently rapidly to preserve filament structures (Ishigami and Hatamo, 1986). The use of detergent extraction of whole amoebae has produced cytoskeletal clusters (at the rear of the cell) which might be focal contacts (*Figure 6*).

(ii) *Strength of cell-substrate adhesion*

Amoebae were allowed to settle in 10 mM NaCl on the wall of a glass capillary tube and a standard volume of solution under investigation was flowed through the tube. A simple measure of strength of cell adhesion was obtained by recording the percentage of cells remaining adherent after the passage of the test solution (King *et al.*, 1982). As the ionic strength of the test solution increased, the percentage of cells remaining adherent increased; in deionized H_2O, 44% of the cells remained adherent but, in 1 mM NaCl, 72% of the cells remained adherent. When a polylysine solution was used almost all the cells remained adherent. Thus there was good correlation between decrease in cell-substrate separation distance (as judged by RIM) and increase in cell adhesion.

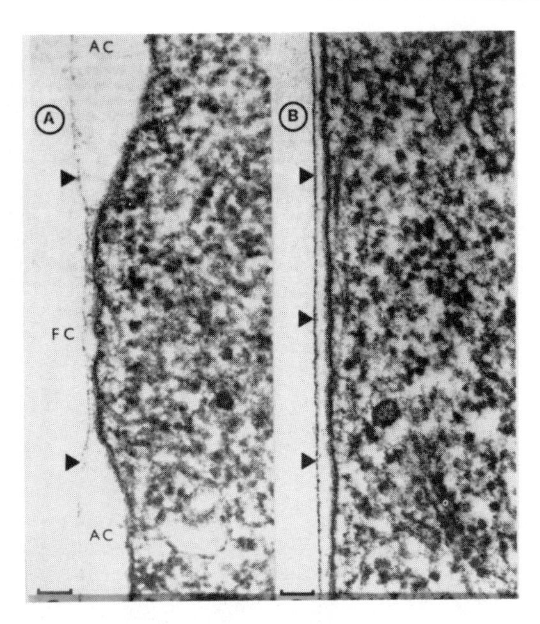

Figure 5. Transmission electron micrographs of cell-substrate gaps from *N. gruberi* amoeba in deionized water moving on two different substrates (from King *et al.*, 1983b). Bar marker = 50 nm. **A**, amoeba fixed while moving on untreated glass. This shows the large cell-substrate gap of the associated contact (AC) and the very close approach of the focal contact (FC). Arrowheads denote the substrate boundary. **B**, Amoeba fixed while moving on polylysine-treated glass showing the uniform close contact.

(iii) *Speed of cell movement*

An increase in the concentration of various solutes (NaCl, KCl, $(NH_4)_2SO_4$, $CaCl_2$) in the medium resulted in increase in motility (King *et al.*, 1979). At 24°C the following motility rates were observed: 0.32 $\mu m\ s^{-1}$ in deionized water, 0.79 $\mu m\ s^{-1}$ in 1 mM NaCl and 1.4 $\mu m\ s^{-1}$ in 10 mM NaCl. Taken with the previous results this supports the hypothesis that decreasing the cell-substrate separation distance results in increased cell adhesion and in turn the development of increased traction on the substrate increased motility. However, it should be pointed out that there is probably an optimum value for cell adhesion above which cell motility is impaired. Results obtained using polylysine solutions produce small cell-substrate separation distances (*Figure 5*) and high cell adhesion but low values for cell motility (King *et al.*, 1982).

The results obtained using RIM provide a dynamic picture of the cell-substrate interactions occurring during amoeboid locomotion (King *et al.*, 1983a). The associated contact changes can reach values of about 10% of the total contact area per second. When the associated contact produced during locomotion in deionized H_2O is compared with that produced in 10 mM NaCl it is evident that the production of new associated contact is much more polarized (i.e. occurring at the anterior end) in the presence of NaCl. *Figure 7* shows the contact pattern over a 4-second period obtained in 10 mM NaCl deionized H_2O.

The newly formed associated contact made at the front end in 10 mM NaCl appears

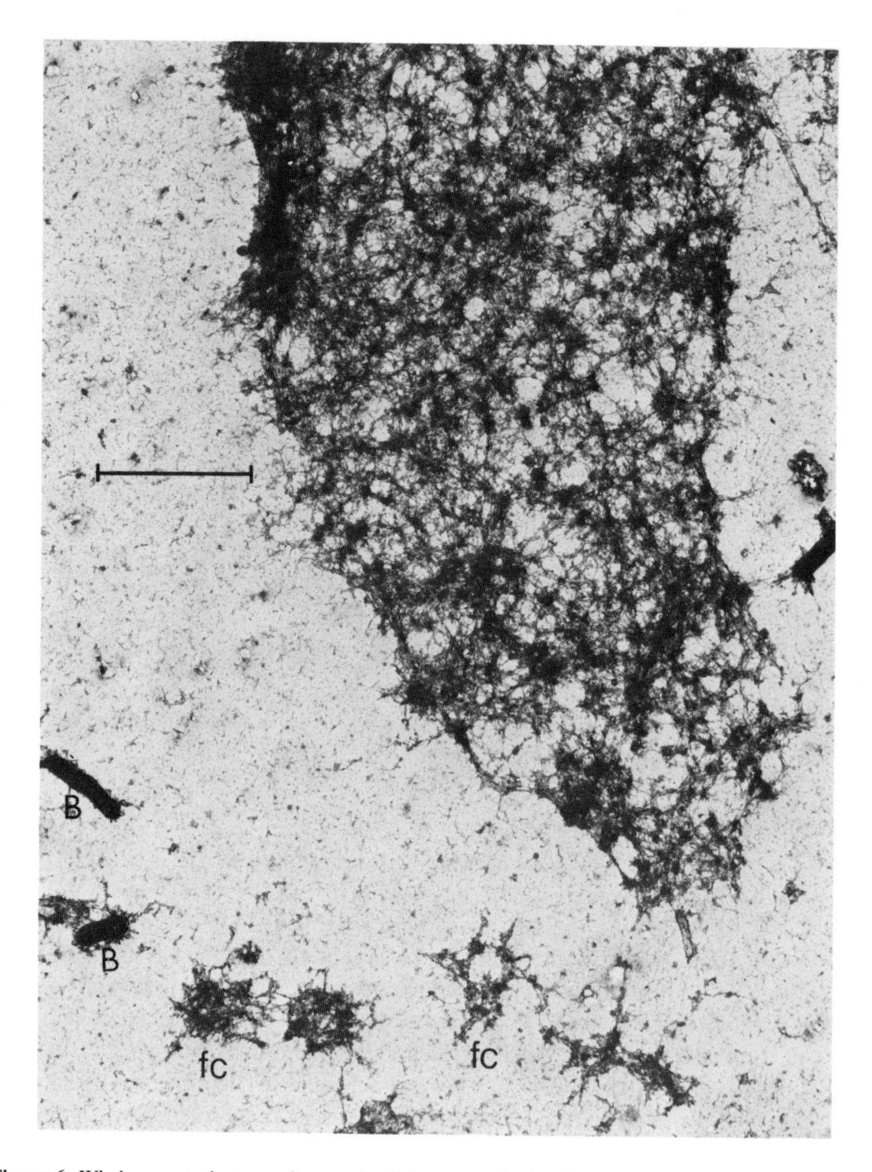

Figure 6. Whole mount electron micrograph of the rear end of a detergent-extracted *N. gruberi* amoeba on a polylysine substrate. fc = presumptive focal contact; B = bacterium; bar marker = 2 μm.

to be more stable than the associated contact formed in deionized H_2O. We believe that focal contacts are much more easily produced when the cell-substrate gap is small and the formation of a focal contact stabilizes the local area of associated contact. In a well-polarized cell, this process will be associated with the eruption of an advancing pseudopodium. One aspect of this study was to consider a 'steering' role for cell-substrate interactions at the front end. The idea of separate 'motor' and 'steering' functions has been considered previously by Grebecki *et al.* (1981), although not with respect to cell-substrate interactions.

Figure 7. Changes in the associated contact pattern of a *N. gruberi* amoeba moving in 10 mM NaCl over a glass substrate at 30°C viewed at 2-s intervals. The horizontally placed bar marker (10 μm) acts as a spatial reference line. The left hand diagram shows the situation at time 0 with the shaded area denoting new associated contact formed in the previous two seconds. The amount of this 'newly formed' contact still present at 2 and 4 s is also shown.

We have not been able to resolve whether there is an increase in focal contact number per associated contact in 10 mM NaCl compared with the number found in deionized water, but there is some evidence that that is the case (King *et al.*, 1979). Nevertheless, one can construct a model system to give an idea of the dynamic nature of focal contact formation.

(i) Consider an amoebae 24 μm long (associated contact = 16 μm) moving at a speed of 0.33 μm s^{-1} in deionized water. Assuming there are, on average, 11 focal contacts within the associated contact area, there will be 11 new focal contacts formed over a migration distance of 16 μm (that is in a period of 48 s). One would naturally expect a similar rate of breakdown at the rear of the cells so that the life expectancy of a single focal contact based on this model would be 48 s.

(ii) Consideration of the same cell moving in 10 mM NaCl at 1.4 μm s^{-1} and assuming the same numbers of focal contacts (11) per associated contact shows that a new focal contact would be formed every second. An alternative model might envisage that the number of focal contacts per associated contact was greater in 10 mM NaCl. If the number were twice as great, then a new contact would be formed every 0.5 s.

Several workers have used adherent latex beads as a probe for studying cell surface behaviour during amoeboid locomotion (Komnick *et al.*, 1973). If a motile *Naegleria* amoeba contacts a latex bead at the front end, then the bead remains in a fixed position relative to the substrate until it reaches the rear of the cell when it moves forward at the same speed as the locomoting amoeba (*Figure 8*).

If several latex beads interact in this way, a cap of beads is formed at the back end of the amoeba. We believe this result is due to the crosslinking of the surface of the bead through the plasma membrane onto the 'stable' cortical tube. In this proposed explanation, there are clear similarities between bead interaction and focal contact formation (*Figure 9*).

This model highlights the concept of formation of a cortical tube at the front end

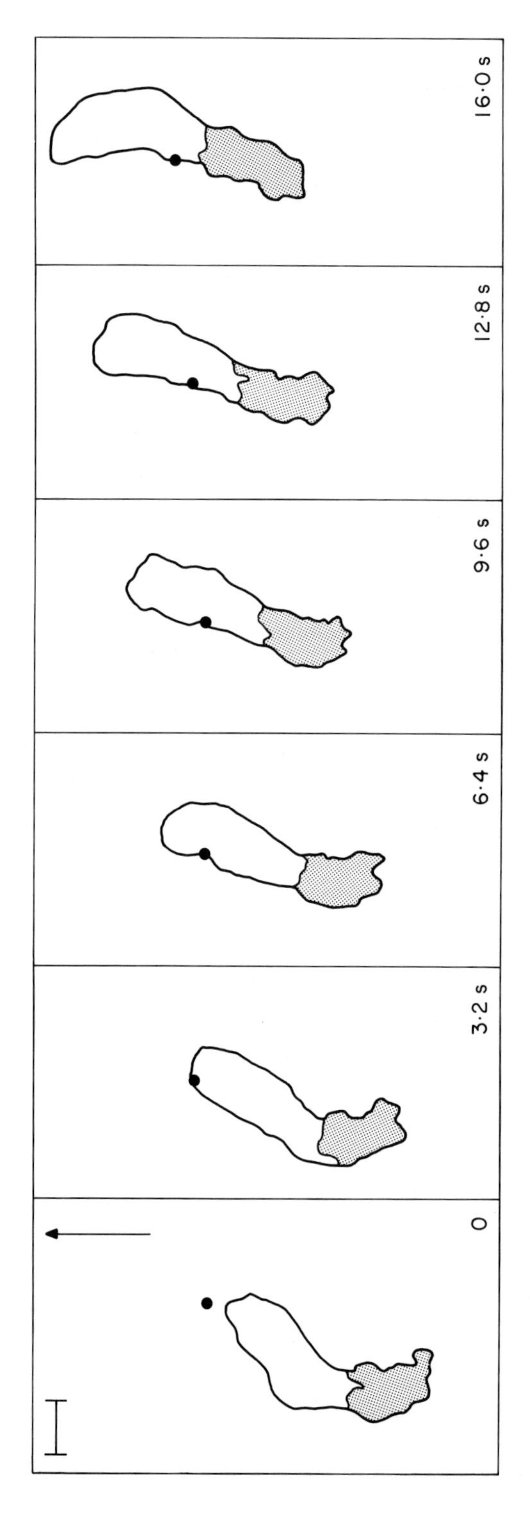

Figure 8. Behaviour of a single 2 μm latex bead (black circle) on interaction with a *N. gruberi* amoeba moving on a glass substrate in 10 mM NaCl. The solid line represents the cell profile of the amoeba and the shaded area of capped latex beads at the posterior of the cell. Bar marker = 10 μm; sequence intervals = 3.2 s.

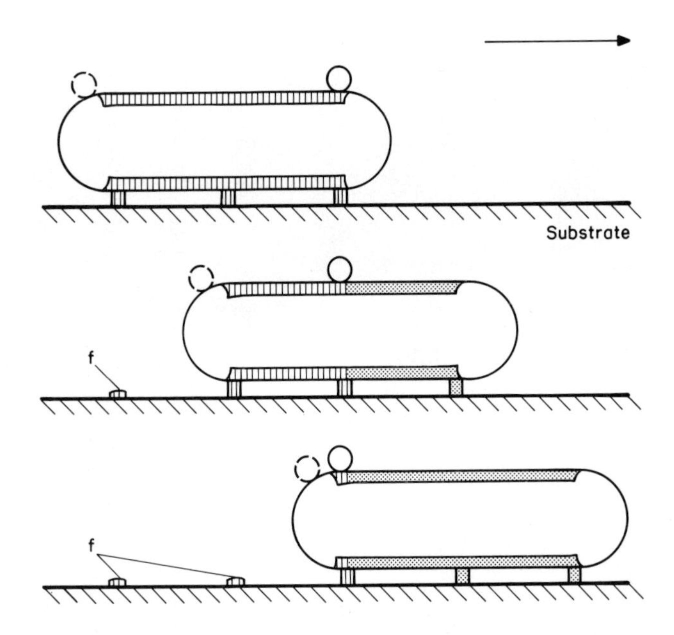

Figure 9. Diagram to show the features of bead interaction, focal contact production and the formation of the cortical tube during locomotion. Assuming the amoeba was moving at 1 μm s^{-1} in 10 mM NaCl and the length of the cortical tube was 16 μm, then the top diagram represents time 0; the middle diagram represents time 8 s and the bottom diagram represents time 16 s. The solid circle represents a latex bead which has recently made contact (i.e. at time 0) and the broken circle represents a single bead present at the rear of the cell (having been previously capped from the front end). The hatched area represents the isotropic meshwork of the cortical tube (extending into the focal contacts at the ventral surface of the amoeba). The dotted shading represents meshwork formed after time 0. The focal contacts (f) left to the rear of the cell may be connected to the cell body by trailing filopodia (not shown).

(sol−gel) and breakdown at the rear end (gel−sol?). One problem in this model is that one might expect the beads to fall off at the rear end due to the collapse of the cortical tube and this clearly does not happen (*Figure 10*). The model might also predict loss of focal contacts at the rear of the cell in a simple manner (in contrast to the results obtained with the latex beads). Using light microscopy, trailing filopodia are sometimes seen to the rear of moving amoebae and we believe that they are formed as a result of the tethering of cytoplasmic extensions to focal contacts which have not broken down. Because they are usually thin structures (less than 0.2 μm) they would escape detection by light microscopy, but in scanning electron microscope studies trailing filopodia are a very common feature (*Figure 10*). Occasionally in the light microscope one can see a trailing filopodium snap and this leads to the recoil of the broken strand into the cell body. The phenomenon of trailing filopodia can be greatly exaggerated by allowing amoebae to locomote on polylysine films where arborescent trails are very well developed and filamentous structures (believed to be microfilaments) are found within the cytoplasm (*Figure 11*).

Occasionally a trailing filopodium can be found that extends back as much as 50 μm behind the cells (i.e. approximately two body lengths). We would expect a cell moving

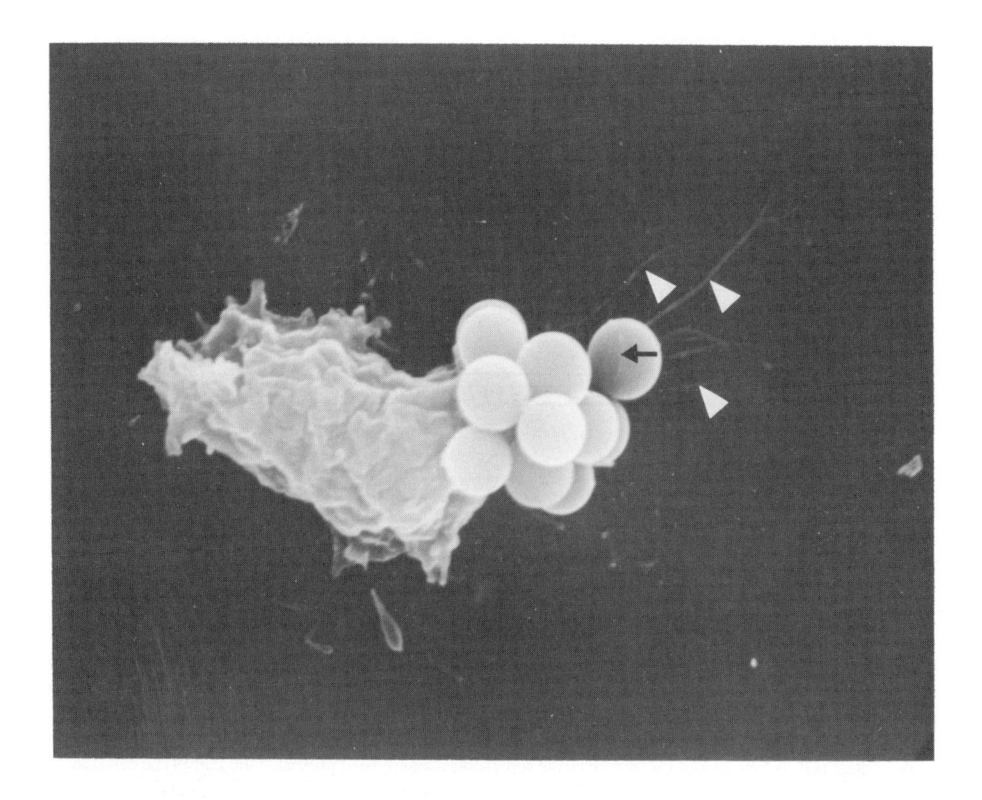

Figure 10. Scanning electron micrograph of a motile *N. gruberi* amoeba which has interacted with 2 μm latex beads in the medium to form a cap of beads at the rear end of the cell. The black arrow shows a protoplasmic filament which terminates just above the arrowhead. The white arrowheads show the trailing filopodia.

at 1 $\mu m\ s^{-1}$ to take 50 s to move this distance and during this time all the focal contact adhesion sites would be broken. These considerations are in marked contrast to the results with beads, where the mass of beads can be carried about for several minutes at the rear end of an amoeba. This gives support to the hypothesis that focal contact and cell body linkage is broken as a result of the development of tension as the cell moves forward.

The cap of beads forms an adhesive mass due to connecting strands of cytoplasmic material (*Figure 10*). At least one focal contact is probably formed on the bead and a 'trailing filopodium' links this to the general cell body (and other beads) at the rear of the cell (*Figure 10*). Thus, events at the rear of the cell during cell movement could be explained by some breakdown in the cortical tube but cytoskeletal elements clearly remain.

Epilogue

There is considerable doubt as to whether a single hypothesis could account for the variety of forms of amoeboid locomotion found in nature (Allen, 1968). Neverthless it could be of value to construct a model based on an amalgam of the features considered

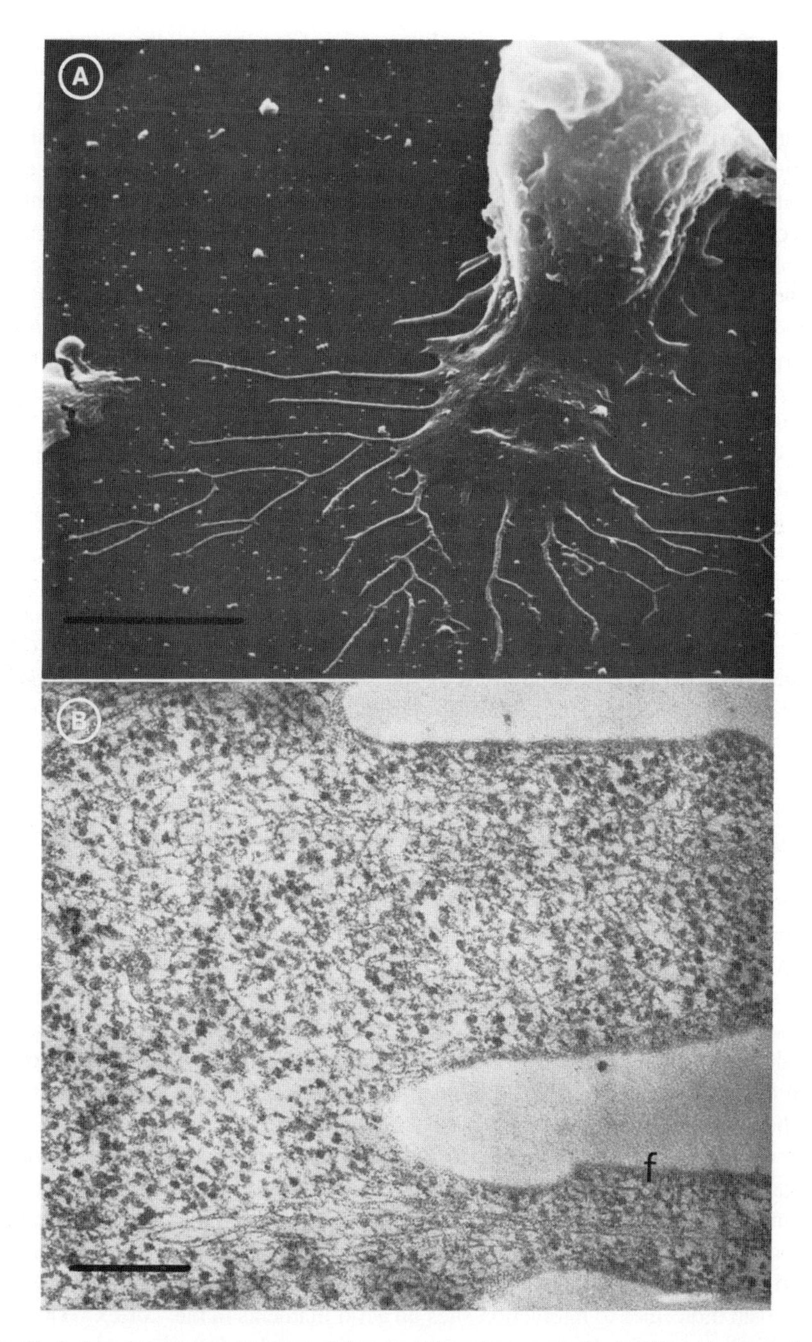

Figure 11. A, Scanning electron micrograph of the rear of *N. gruberi* amoeba fixed while moving on a polylysine-coated substrate. The trailing filopodia are clearly visible to the rear of the cell body. Bar marker = 5 μm. **B**, Transmission electron micrograph through a trail similar to that shown in **A.** Organized bundles of filaments can be seen in the core of the trailing filopodium (f) and they extend into the flattened zone of the cell body. Bar marker = 0.2 μm. (Reproduced by courtesy of Bob Miller.)

for *Dictyostelium* and *Naegleria*, involving microfilaments in the generation of cytoskeletal and cytomuscular elements during locomotion with respect to the anterior–posterior axis of the cell. Four facets need to be integrated in order that crawling polarized movement occurs – formation and breakdown of the cortical tube, generation of traction on the substrate, regulation of actin–myosin contractile events and production of the forward projections (for example pseudopodia) at the front end.

The cortical tube can be envisaged as a cross-linked meshwork formed by interactions of microfilaments with gelation factors (for example the 120 K protein in *Dictyostelium*). Formation at the anterior of the cell would have to proceed at a rate commensurate with the locomotion rate. Thus an amoeba moving at 2 μm s^{-1} would need to form new cortical tube at this rate. This might appear to be a high value but rates of up to 10 μm s^{-1} have been observed for polymerization rates of actin to produce microfilaments during the acrosome reaction of *Thyone* sperm (Tilney *et al.*, 1981). It is also possible that the initial building units for the MF might be oligomers of actin rather than G-actin itself. The proposed rigidity of the cortical meshwork can be most easily appreciated by reference to the quotation from Wolesewick and Condeelis (1986) – 'The cortical network could function to freeze cell shape and surface contour inhibit lateral mobility of components in the plane of the membrane and prevent incursion into the cortex of large organelles from the cytoplasm'. At the ventral surface, the associated contact platform initially forms as a result of electrodynamic forces of attraction (Curtis, 1964) but this soon becomes rigidified by the cortical meshwork formation. *Figure 7* shows the relatively stable character of the associated contact in 10 mM NaCl medium prior to breakdown at the rear of the cell.

In order to develop traction on the substrate it is proposed that the cytoskeletal elements of the focal contacts and associated subtending filopodia insert into the cortical meshwork of the cell body and therefore the cytoskeleton is locked onto the substrate providing the grip necessary for forward movement to occur. Support for this proposal comes from our work on the amoeboid locomotion of *Acanthamoeba* (Preston and King, 1984). This genus derives its name from the fact that the surface of the amoeba is covered with spines (acanthapodia). Each acanthapodium possesses a central core of microfilaments and may reach dimensions of 10 μm long × 2 μm in diameter. The spine-like acanthopodia are formed at the front end of the cell and can conveniently be visualized at the lateral margins of the cell by light microscopy. An individual acanthapodium once formed remains fixed in position with respect to the substrate (although unattached to the substrate) until it breaks down towards the rear of the cell.

In the case of *Dictyostelium* there is a considerable amount of experimental evidence to suggest that the contraction event occurs in the posterior region of the cell via actin–myosin interactions. There is also evidence to suggest that myosin is required in the polymeric form (Reines and Clarke, 1985). In order for contraction to occur the rigid cortical meshwork must be broken down to some extent to allow entry of the large myosin molecules to interaction sites on actin filaments in the 'cortex'. The group of actin-binding proteins known as severing proteins might play a key role as solation factors, that is factors which promote the change from gel cytoplasm to sol cytoplasm. The role of these solation factors would be 2-fold – breakdown of the rigid cortical tube and thereby allowing a more open cytoskeletal structure to develop in which contraction events could occur. A key function of the contraction event would be to move cytoplasmic components forward to provide material for the advancing pseudopodium.

The eruption of a pseudopodium or the more gradual advance of a sheet-like lamellipodium at the front end of a crawling cell requires some degree of plasticity in contrast to the rigidity proposed for the cortical tube. A common feature of these areas is their hyaline character, that is they are clear zones excluding intracellular organelles. The hyaline area in the giant amoeba was found to be very watery but the lamellipodia and pseudopodia of *Dictyostelium* amoeba are filled with dense filament networks (Wolesewick and Condeelis, 1986). It appears likely that the addition of actin subunits to the membrane associated ends of existing microfilaments could account for protrusive activity in a similar way to that described for the lamellipodia of motile fibroblasts (Wang, 1985).

Acknowledgements

We would like to thank Terry Preston for helpful discussion, Lynda King for the artwork and Linda McCracken for typing the manuscript.

References

Alberts,B., Bray,D., Lewis,J., Raff,M., Roberts,K. and Watson,J. (1983) The cytoskeleton. In *Molecular Biology of the Cell*. Garland Publishing, New York, pp. 550–610.

Allen,N.S. and Allen,R.D. (1978) Cytoplasmic streaming in green plants. *Annual Reviews of Biophysics and Bioengineering,* **7**, 469–495.

Allen,R.D. (1961) A new theory of amoeboid locomotion and protoplasmic streaming. *Experimental Cell Research,* Supplement **8**, 17–31.

Allen,R.D. (1968) Differences of a fundamental nature among several types of amoeboid movement. *Symposium of the Society of Experimental Biology*, **22**, 151–168.

Bennett,T.H. and Condeelis,J. (1984) Decoration with myosin sub fragment-1 disrupts contacts between microfilaments and the cell membrane in isolated *Dictyostelium* cortices. *Journal of Cell Biology,* **99**, 1434–1440.

Condeelis,J. (1981) Microfilament–membrane interactions in cell shape and cell architecture. In *International Cell Biology 1980–1981.* Schweiger,H.G. (ed.), Springer-Verlag, Berlin, pp. 306–320.

Condeelis,J., Vahey,M., Carboni,J.M., Demey,J. and Ogihara,S. (1984) Properties of the 120 000 and 95 000-dalton actin-binding proteins from *Dictyostelium discoideum* and the possible functions in assembling the cytoplasmic matrix. *Journal of Cell Biology,* **99**, 119s–126s.

Curtis,A.S.G. (1964) Mechanism of adhesion of cells to glass, a study by interference reflection microscopy. *Journal of Cell Biology,* **20**, 199–215.

Edgar,L.A.and Zavortink,M. (1983) The mechanism of diatom locomotion. II. Identification of actin. *Proceedings of the Royal Society of London Series B,* **218**, 345–348.

Fukui,Y. and Yumura,S. (1986) Actomyosin dynamics in chemotactic amoeboid movement of *Dictyostelium. Cell Motility and the Cytoskeleton*, **6**, 662–673.

Grebecka,L. and Grebecki,A. (1981) Testing the motor functions of the frontal zone in the locomotion of *Amoeba proteus. Cell Biology International Reports,* **5**, 587–594.

Grebecki,A., Grebecka,L. and Klopocka,W. (1981) Testing steering functions of the frontal zone in the locomotion of *Amoeba proteus. Cell Biology International Reports,* **5**, 595–600.

Hellewell,S.B. and Taylor,D.L. (1979) The contractile basis of amoeboid movement. IV. The solation–contraction coupling hypothesis. *Journal of Cell Biology,* **83**, 633–648.

Hoffmann-Berling,H. (1956) Das Kontraktile eiweiss undifferenzierten zellen. *Biochimica Biophysica Acta,* **191**, 453–463.

Ishigami,M. and Hatano,S. (1986) Dynamic aspects of the contractile system in *Physarum* plasmodium. II. *Cell Motility and the Cytoskeleton,* **6**, 448–457.

John,D.T. (1982) Primary amebic meningoencephalitis and the biology of *Naegleria fowleri*. *Annual Reviews of Microbiology,* **36**, 101–123.

King,C.A., Westwood,R., Cooper,L. and Preston,T.M. (1979) Speed of locomotion of the soil amoeba *Naegleria gruberi* in media of different ionic compositions with special reference to interactions with the substratum. *Protoplasma,* **99**, 323–334.

King,C.A., Preston,T.M., Miller,R.M. and Grose,C. (1982) The cell surface in amoeboid locomotion – studies on the role of cell–substrate adhesion. *Cell Biology International Reports,* **6**, 893–900.

King,C.A., Cooper,L. and Preston,T.M. (1983a) Cell–substrate interactions during amoeboid locomotion of *Naegleria gruberi* with special reference to alterations in temperature and electrolyte concentration. *Protoplasma,* **118**, 10–18.

King,C.A., Preston,T.M. and Miller,R.M. (1983b) Cell–substrate interactions in amoeboid locomotion – a matched reflexion interference and transmission electron microscopy study. *Cell Biology International Reports,* **7**, 641–649.

King,C.A., Whitehead,C., Pringle,N., Cooper,L. and Baines,I.C. (1986) Motility of protozoan gregarines – a model for studying mechanochemical force transduction at the cell surface. *Cell Motility and the Cytoskeleton,* **6**, 243.

Komnick,H., Stockem,W. and Wohlfarth-Bottermann,K.E. (1973) Cell mobility: mechanisms in protoplasmic streaming and amoeboid movement. *International Review of Cytology,* **34**, 169–249.

Lee,J.J., Hutner,S.H. and Bovee,E.C. (1985) *An Illustrated Guide to the Protozoa. Society of Protozoologists,* Kansas City, pp. 629.

McRobbie,S.J. (1986) Chemotaxis and cell motility in the cellular slime molds. *CRC Critical Reviews in Microbiology,* **13**, 335–375.

Mast,S.O. (1926) Structure movement, locomotion and stimulation in amoeba. *Journal of Morphology and Physiology,* **41**, 347–425.

Pollard,T.D. (1981) Cytoplasmic contractile proteins. *Journal of Cell Biology,* **91**, 156s–165s.

Pollard,T.D. and Cooper,J.A. (1986) Actin and actin-binding proteins. A critical evaluation of mechanisms and functions. *Annual Reviews of Biochemistry,* **55**, 987–1035.

Preston,T.M. (1985) A prominent microtubule cytoskeleton in *Aanthamoeba. Cell Biology International Reports,* **9**, 307–314.

Preston,T.M. and King,C.A. (1978) An experimental study of the interaction between the soil amoeba *Naegleria gruberi* and a glass substrate during amoeboid locomotion. *Journal of Cell Science,* **34**, 145–158.

Preston,T.M. and King,C.A. (1984) Amoeboid locomotion of *Acanthamoeba castellanii* with special references to cell-substratum interactions. *Journal of General Microbiology,* **130**, 2317–2323.

Reines,D. and Clarke,M. (1985) Immunochemical analysis of the supramolecular structure of myosin in contractile cytoskeletons of *Dictyostelium* amoebae. *Journal of Biological Chemistry,* **260**, 14248–14254.

Russell,D.G. and Sinden,R.E. (1981) Role of the cytoskeleton in the motility of coccidian sporozoites. *Journal of Cell Science,* **50**, 345–359.

Sheetz,M.P. and Spudich,J.A. (1983) Movement of myosin-coated fluorescent beads on actin cables *in vitro*. *Nature,* **303**, 31–35.

Stockem,W., Weber,K. and Wehland,J. (1978) The influence of microinjected phalloidin on locomotion, protoplasmic streaming and cytoplasmic organisation in *Amoeba proteus* and *Physarum polycephalum*. *Cytobiologie,* **18**, 114–131.

Taylor,D.L. and Condeelis,J.S. (1979) Cytoplasmic structure and contractility in amoeboid cells. *International Review of Cytology,* **56**, 57–144.

Tilney,L.G., Bonder,E.M. and de Rosier,D.J. (1981) Actin filaments elongate from their membrane-associated ends. *Journal of Cell Biology,* **90**, 485–494.

Wang,Y.-L. (1985) Exchange of actin subunits at the leading edge of living fibroblasts: possible role of treadmilling. *Journal of Cell Biology,* **101**, 597–602.

Wehland,J., Weber,K., Gawlitta,W. and Stockem,W. (1979) Effects of the actin-binding protein DNase I on cytoplasmic streaming and ultrastructure of *Amoeba proteus. Cell and Tissue Research,* **199**, 353–372.

Wolosewick,J.J. and Condeelis,J. (1986) Fine structure of gels prepared from an actin-binding protein and actin; comparison to cytoplasmic extracts and cortical cytoplasm in amoeboid cells *Dictyostelium discoideum. Journal of Cellular Biochemistry,* **30**, 227–243.

Yumura,S. and Fukui,Y. (1985) Reversible cyclic AMP-dependent changes in distribution of myosin thick filaments in *Dictyostelium. Nature,* **314**, 194–196.

CHAPTER 8

Spatial association of F-actin with growth polarity and septation in the fission yeast *Schizosaccharomyces pombe*

J.MARKS, I.M.HAGAN and J.S.HYAMS

Department of Botany and Microbiology, University College London, Gower Street, London WC1E 6BT, UK

Introduction

The fission yeast *Schizosaccharomyces pombe* grows only by elongation at its poles, maintaining a constant diameter throughout the cell division cycle (Mitchison and Nurse, 1985). Underlying this superficially simple mode of growth, however, is a sequence of precisely orchestrated switches in the sites of cell wall deposition. These can be summarized as follows:

(i) following separation, daughter cells grow only at the end that existed prior to cytokinesis (the old end);

(ii) monopolar growth is maintained for one third of the cell cycle at which point growth at the new end (the product of cytokinesis) is initiated and cells thereafter grow in a bipolar manner;

(iii) approximately three quarters of the way through the cell cycle, end growth ceases and the cell deposits a septum and undergoes cytokinesis.

Despite the asymmetric nature of this pattern of growth, with both monopolar and bipolar extension, cells show a remarkable fidelity in the equatorial positioning of the nucleus and, transiently, the septum. It is the involvement of the cytoskeletal proteins, primarily actin, in the determination of growth polarity and nuclear and septal positioning that we address in this study. In particular, we have utilized techniques to localize elements of the yeast cytoskeleton (Adams and Pringle, 1984; Kilmartin and Adams, 1984), cell wall (Darken, 1961; Maeda and Ishida, 1967) and nucleus (Williamson and Fennell, 1975) in the fluorescence microscope to catalogue structural rearrangements of cytoskeletal proteins through the yeast cell cycle and to correlate these with the patterns of wall growth and mitosis (Marks and Hyams, 1985; Marks *et al.*, 1986). The availability of a number of temperature-sensitive (ts), cell division cycle (*cdc*) mutants of *S. pombe*, which normally arrest around the time of mitosis and cytokinesis (Nurse *et al.*, 1976), have also proved invaluable in confirming and furthering our initial observations on wild type cells.

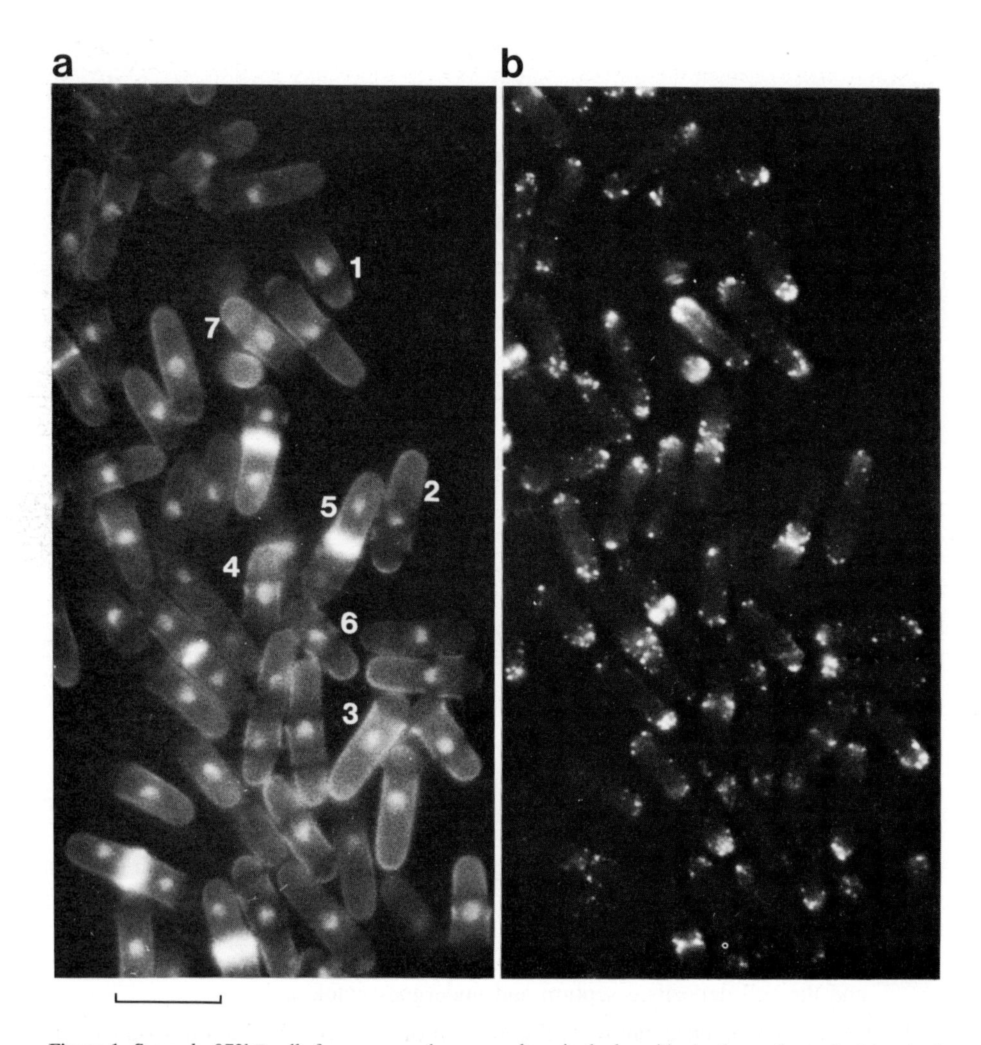

Figure 1. *S. pombe* 972h$^-$ cells from an asynchronous culture in the logarithmic phase of growth; (**a**) stained with calcofluor and DAPI, (**b**) stained with rhodamine-conjugated phalloidin. Scale bar = 10 μm.

Structural rearrangements of actin through the cell division cycle

Switches in the pattern of cell wall deposition in *S. pombe* from monopolar to bipolar and from bipolar to equatorial generates areas of cell wall having architectural and compositional differences. These vary in their degree of fluorescence when stained with 'brighteners' such as calcofluor (Darken, 1961), and thus are useful markers of progress through the cell division cycle (Streiblova, 1984). The growth characteristics of *S. pombe* vary from strain to strain and are further influenced by factors such as composition of the growth medium, oxygen tension and temperature (Mitchison, 1970; Johnson and McDonald, 1983). Inconsistencies in experimental technique between different groups has no doubt been responsible for the dispute that exists over the precise mode of elongation (Mitchison, 1957; May, 1962; Streiblova *et al.*, 1966; Johnson, 1968; Biely

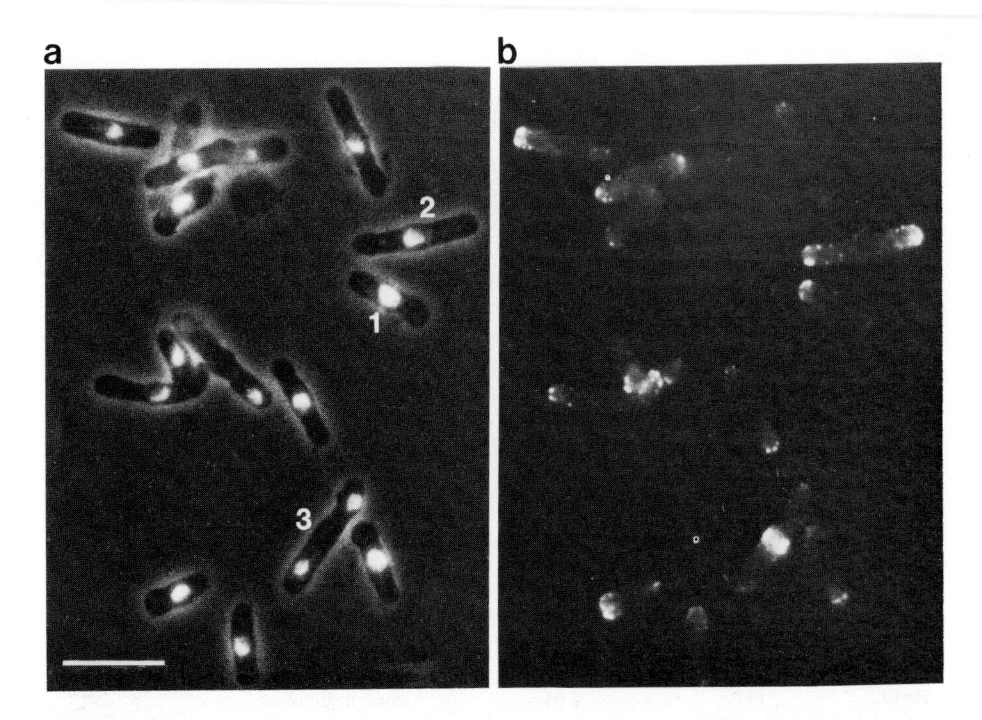

Figure 2. Cells were cultured as in *Figure 1*: (**a**) nuclei stained with DAPI, and cell outlines shown using phase-contrast, (**b**) actin visualized with an anti-actin monoclonal antibody. Scale bar = 10 μm.

et al., 1973; Mitchison and Nurse, 1985; Miyata *et al.*, 1986). The most complete study is that of Mitchison and Nurse (1985), and by accurately reproducing the conditions they employed we are able to correlate our observations with their analysis of growth through the cell cycle. Following division, daughter cells inherit one end from their mother (the old end) and the other (the new end) from the preceding septation event (Mitchison, 1957; May, 1962; Johnson, 1965; Mitchison and Nurse, 1985). The old end stains intensely with calcofluor but the new end shows little fluorescence and appears as a dark hemisphere (*Figure 1a,* cell 1). Cells of the wild type strain 972h$^-$ showing a single brightly staining end can be identified as being before 0.34 of the cell cycle (Mitchison and Nurse, 1985). As growth at the new end is initiated (a point designated 'NETO', for new end take off), the dimly staining region becomes internalized forming the 'birth scar' (*Figure 1a*, cell 2). This identifies cells as being between 0.34 and 0.75 of the cell cycle. Cells in the remaining portion of the cycle possess a brightly staining septum (*Figure 1a*, cell 5). A more precise idea of the position of the cell within these three categories can be gained by combining the above criteria with those of cell length, position in mitosis and intensity of septum staining.

Wild type cells from a population in the logarithmic phase of growth, stained with two specific probes for F-actin, rhodamine-conjugated phalloidin and a monoclonal anti-actin antibody, are shown in *Figures 1b* and *2b* respectively, with their corresponding calcofluor/DAPI (4′,6-diamidino-2-phenylindole) or phase/DAPI images in *Figures 1a* and *2a*. Correlation of these staining patterns reveals an interesting and unambiguous

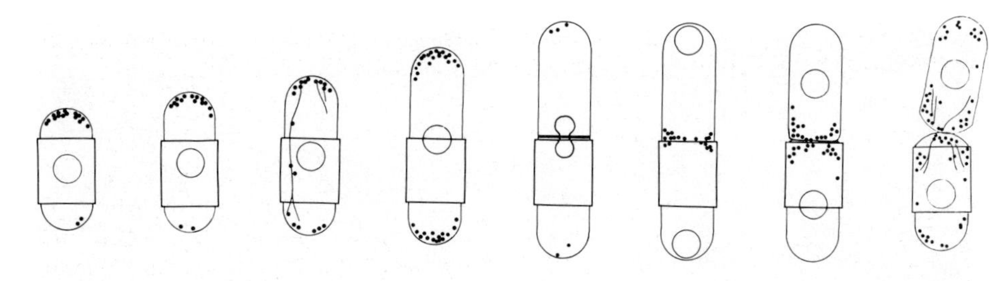

Figure 3. Schematic representation of the rearrangements of actin dots and filaments through the *S. pombe* cell division cycle.

Table I. Summary of the characteristics of selected *cdc* mutants .

Gene	*Allele*	*Transition point*[a]	*Defect*	*Reference*[b]
cdc 3	124	0.73	Late septation	1
cdc 4	8	0.80	Late septation	1
cdc 10	129	−0.10	DNA synthesis	1
cdc 11	136	0.81	Early septation	1
cdc 12	112	0.82	Late septation	1
cdc 13	117	0.64	Nuclear division	1
cdc 16	116	0.82	Cell separation	2
ben 4	D3	0.78	Mitosis	3

[a]Point in the cell cycle at which the temperature-sensitive function has been completed.
[b]Reference: (1) Nurse *et al.* (1976); (2) Minet *et al.* (1979); (3) Roy and Fantes (1982).

temporal and spatial association between sites of wall deposition and localization of F-actin.

In cells identified either by their length or by calcofluor staining as exhibiting single end growth (cell 1 in *Figures 1* and *2*), F-actin is visualized as an array of 'dots' or 'patches' associated predominantly with the growing tip. As bipolar growth begins, with the consequent internalization of the birth scar and bright calcofluor staining at both cell poles, F-actin shows a corresponding bipolar distribution (*Figures 1* and *2*, cell 2 in *Figure 1a*). At a cell length of about 14 μm, [the so-called 'constant volume stage' (Mitchison, 1957)], growth ceases as the cells prepare for division. Concomitantly actin almost totally disappears from the cell ends and relocalizes as a filamentous 'ring' or 'belt' at the equator. Cell 3 in *Figure 1* shows the earliest stage in this transition. A few dots remain at the poles but a feint equatorial ring, which overlies the nucleus, has appeared and persists throughout anaphase (arrow in *Figure 1b*). The appearance of the ring precedes both the formation of the septum and detectable nuclear division. As anaphase proceeds, the ring increases in intensity (cell 4 in *Figure 1*) and, towards the end of mitosis, septum formation begins (cell 5, *Figure 1*, see also cell 3, *Figure 2*). As the septum consolidates equatorial actin becomes less discrete and appears as arrays of dots on either side of it (*Figure 1a*, cell 5). The final gross rearrangement of both the site of wall growth and the position of F-actin occurs towards the end of

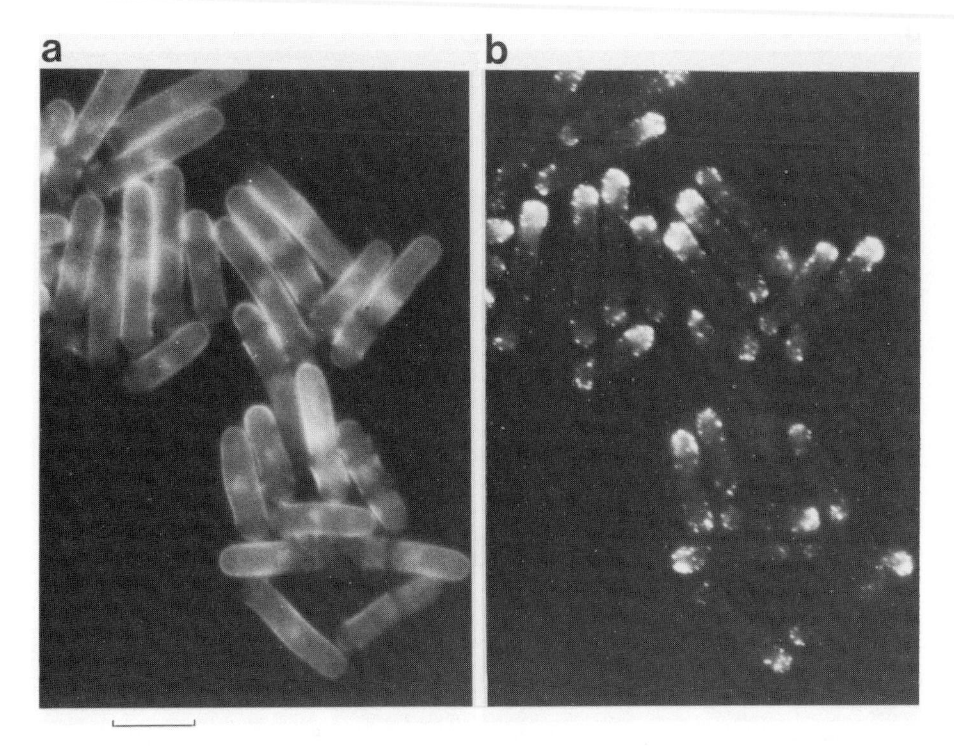

Figure 4. Temperature-arrested *cdc* 10 cells incubated for 6 h at 36°C; (**a**) stained with calcofluor, (**b**) stained with rhodamine-conjugated phalloidin. Note the predominance of growth at a single pole with minor extension (visualized as a cap of calcofluor staining) at the other. The density of actin shows a corresponding asymmetric distribution. Scale bar = 10 μm.

cytokinesis when sister cells physically separate. At this stage actin is still localized at the incipient new ends (i.e. those recently exposed by cytokinesis), although the initial growth of the siblings will be at the opposite, old end. Accordingly, a rapid deployment of F-actin to the growing poles occurs early in the ensuing cell cycle.

In addition to dots, F-actin filaments are also visualized in some cells. These invariably have one end associated with an array of dots, either at the cell poles or equator, the other being free in the cytoplasm and associated with no obvious structure. Filaments with both ends terminating in the cytoplasm are never seen. Interestingly, filaments are seen most commonly at points in the cell cycle characterized by a gross rearrangement of the dots such as the initiation of bipolar growth (*Figure 1a*, cells 6 and 7), and preceding the formation of the septum (not shown). These events are summarized in *Figure 3*.

Analysis of F-actin localization in *cdc* mutants

Further evidence for the relationship between actin and cell growth and division in fission yeast has been obtained from various ts *cdc* mutants. These become arrested at specific points in the cell division cycle when transferred to the restrictive temperature (Bonatti *et al.*, 1972; Nurse *et al.*, 1976). Using rhodamine-conjugated phalloidin, we have

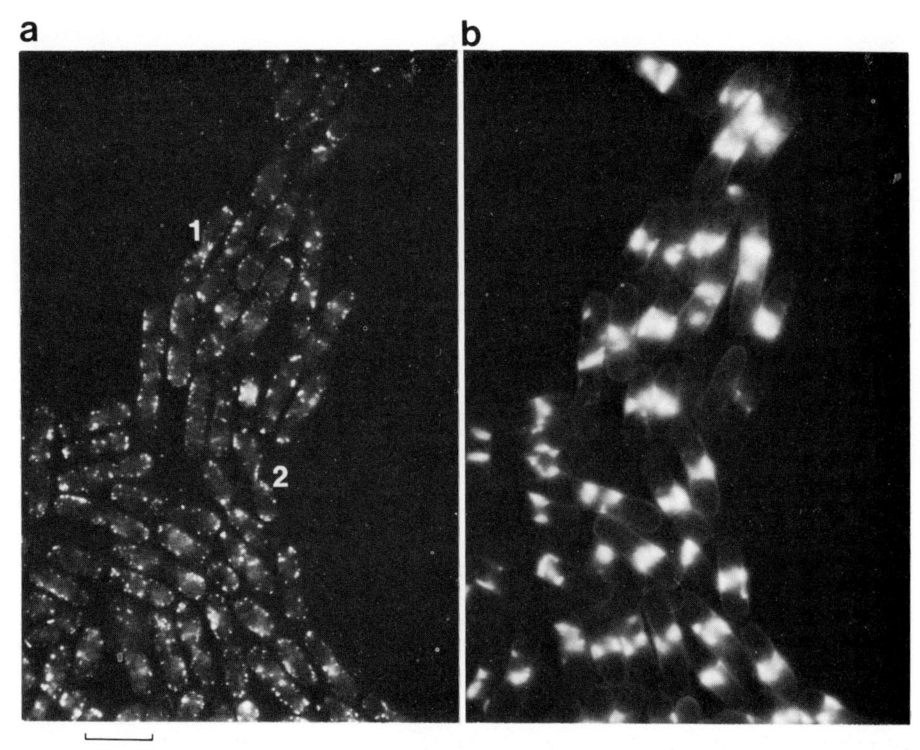

Figure 5. Temperature-arrested *cdc* 3 cells incubated for 8 h at 36°C; (**a**) stained with rhodamine-conjugated phalloidin, (**b**) stained with calcofluor. Absence of polar growth and defective Septation are associated with delocalized actin. Scale bar = 10 μm.

investigated the distribution of F-actin and its association with cell wall deposition in a number of *cdc* mutants. We have also examined the cold-sensitive, benomyl-resistant mutant *ben* 4 (Roy and Fantes, 1982). Some of the properties of these mutants are summarized in *Table 1*.

Actin at the new end before 'new end take off'

Phalloidin-stained wild type cells invariably show a small amount of fluorescence at the 'non-growing' end in the first third of the cell cycle. We have investigated the possibility that a minor amount of wall deposition at the new end might contribute to growth during this period using the mutant *cdc* 10, which blocks in G1 (Nurse *et al.*, 1976). The logic of this experiment is that events that might be barely detectable in wild type cells would be exaggerated in cells forced by temperature arrest to spend an extended period in this portion of the cell cycle. Calcofluor and phalloidin images of *cdc* 10 are shown in *Figure 4*. As expected, fluorescence is predominantly at the growing (old) end although a lesser number of actin dots and a corresponding cap of wall fluorescence are always seen at the new end.

Distribution of actin in septation mutants

Mutants *cdc* 3 and *cdc* 12 have been characterized as late septation mutants (Nurse *et al.*, 1976), on the criterion that, although they deposit cell wall material, it is

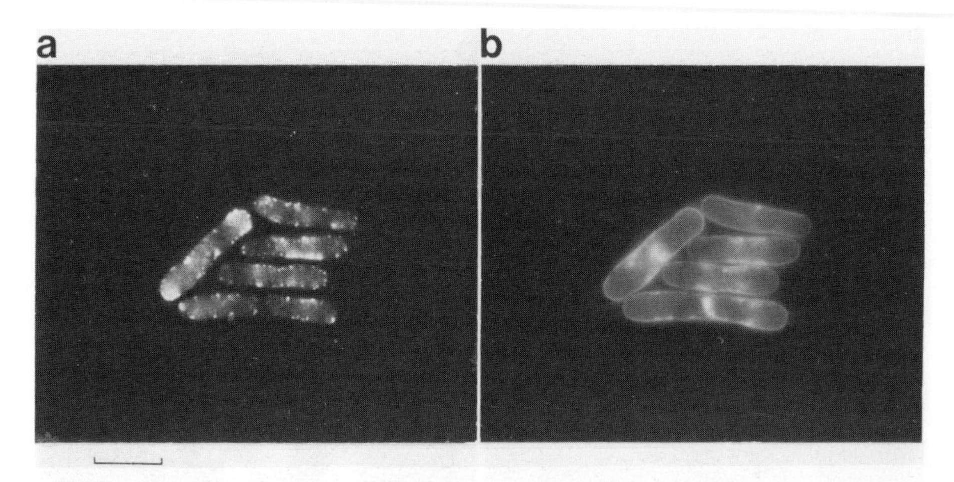

Figure 6. Temperature-arrested *cdc* 12 cells incubated for 6 h at 36°C; (**a**) stained with rhodamine-conjugated phalloidin, (**b**) stained with calcofluor. As with *cdc* 3, actin is delocalized and primarily associated with the cell cortex. Cells show a corresponding absence of polar growth, becoming swollen at their tips and failing to complete cytokinesis. Scale bar = 10 μm.

disorganized and the cell fails to form a functional barrier across its cytoplasm (Streiblova *et al.*, 1984). These strains reveal essentially similar patterns of cell wall and F-actin distribution (*Figures 5* and *6*). Cells arrest at the restrictive temperature with a dumb-bell shape rather than the typical cylindrical morphology of wild type cells and show little or no elongation. Both mutants deposit distinct areas of fluorescent wall material at and around the cell equator, although in *cdc* 12 this does not appear to extend as far into the cytoplasm as in *cdc* 3, in which consolidating cell wall material is sometimes seen to result in the constriction of the cytoplasm (*Figure 5*, cells 1 and 2). These mutants also show a pattern of F-actin distribution markedly different to that seen in wild type cells with dots located predominantly at the cell cortex and with no obvious polar bias. In *cdc* 12, polar actin is sometimes visualized following a 6 h temperature block but, as in *cdc* 3, cortical actin prevails. In both, however, concentrations of F-actin are clearly associated with localized deposits of cell wall. These mutants appear to lack the filaments that typify certain cell cycle transitions in wild type cells, although the significance of this is not clear.

Actin distribution in multiseptate mutants

A different kind of cytokinetic defect is exhibited by *cdc* 4, *cdc* 13 and *cdc* 16, which initiate multiple 'septa' but fail to complete cytokinesis when grown at 36°C. Mutant *cdc* 16 does not show marked elongation but rather swells to produce dumb-bell shapes once cells have reached a length of about 13 μm, whereupon septum production is initiated (Minet *et al.*, 1979). An example of such cells is given in *Figure 7*; arrested cells have the bulk of their F-actin associated with those regions corresponding to areas of 'septum' deposition. There is a paucity of both polar staining and of F-actin filaments in these cells, although, where a linear region of cell wall material arises tangentially from an aberrant septum, it is associated with a bar of actin (arrowed, *Figure 7*). It is noteworthy that F-actin can be visualized in both the end and middle compartments

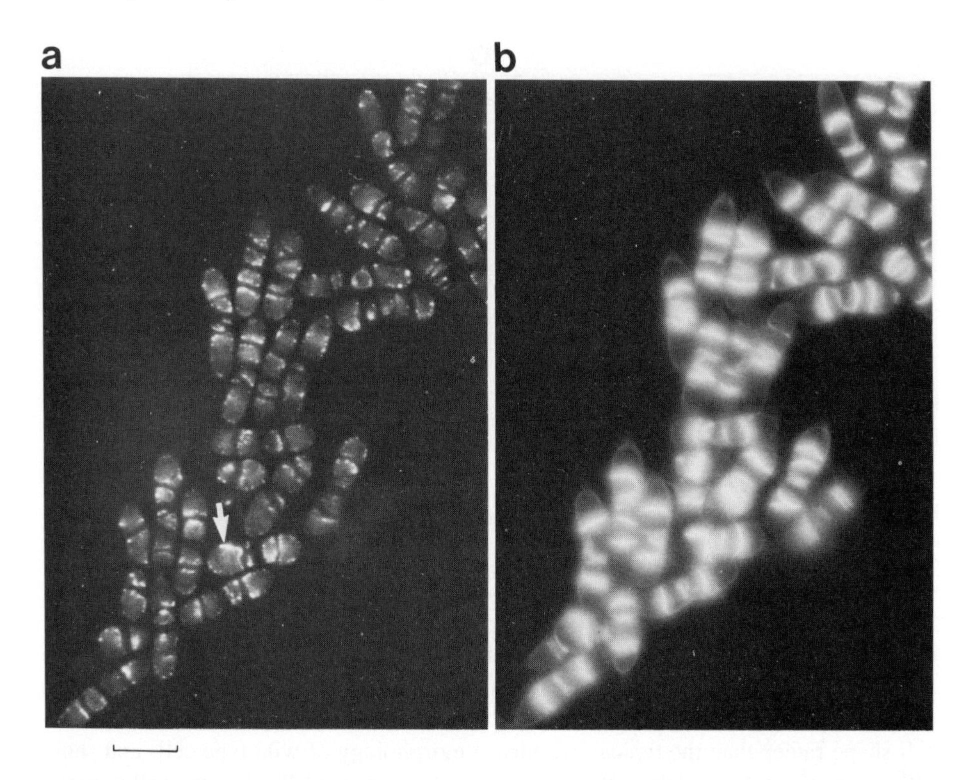

Figure 7. Multiseptate temperature-arrested *cdc* 16 cells blocked for 8 h at 36°C; (**a**) stained with rhodamine-conjugated phalloidin, (**b**) stained with calcofluor. In this strain, actin is primarily associated with those septa furthest from the cell equator, and also underlies 'bars' of delocalized cell wall material (see arrowed cell in **a**). Scale bar = 10 μm.

of the cell but is biased in its distribution towards the septa furthest from the cell equator, which are presumably the most recently formed.

Mutant *cdc* 4 also shows the formation of multiple 'septa' but these are more variable in appearance than those produced by *cdc* 16. *cdc* 4 continues to grow at the restrictive temperature and, consistent with this, actin dots are often localized at the cell poles (*Figure 8*, cell 1). Actin is also coincident with the 'septa', but, as with *cdc* 16 is most conspicuously associated with those flanking the middle or original septum (e.g. *Figure 8*, cells 2 and 3). The most obvious and unequivocal association between F-actin and calcofluor-staining material is again seen where elements of wall material trail off into the cytoplasm and follow the same contours as rhodamine–phalloidin-staining structures (e.g. *Figure 8*, cell 4).

Mutant *cdc* 13 is not a septation mutant as such but rather an atypical nuclear division mutant (Nasmyth and Nurse, 1981) which, unlike the majority of mutants in this class, goes on to form 'septa'. This is shown clearly in *Figure 9* where, following a 7 h temperature block, cells have undergone considerable elongation. F-actin is abundant at the growing poles, coincident with bright calcofluor staining (cells 1 and 2; actin filaments are also particularly obvious in these cells (cells 1, 2 and 4). End staining declines with the appearance of a filamentous ring towards the middle of the cell

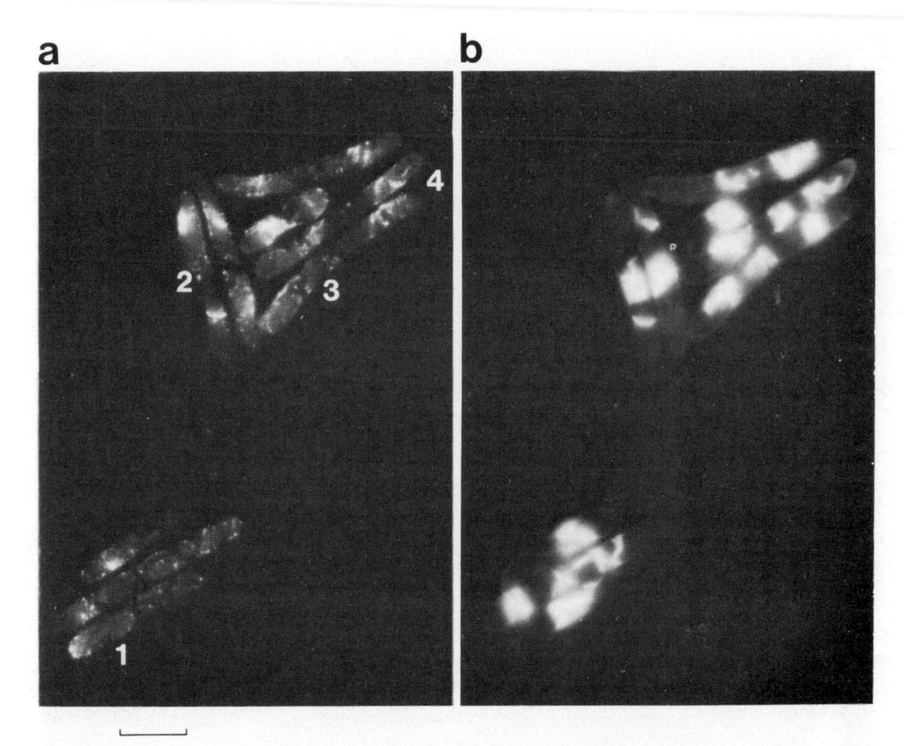

Figure 8. *cdc* 4 cells temperature-arrested at 36°C for 8 h; (**a**) stained with rhodamine-conjugated phalloidin, (**b**) stained wtih calcofluor. As with *cdc* 16, these cells exhibit actin associated largely with those septa furthest from the cell equator (presumably those most recently formed), and with delocalized extrusions of cell wall material. Scale bar = 10 μm.

presumably predicting the impending site of septation (cells 3 and 4). In cells with one or two mature septa (cells 1 and 2), there is typically a feint fluorescence associated with actin dots, which appear, as in wild type cells, to be relocating away from the septum at this stage. In cells forming multiple septa, these clearly co-localize with larger diffuse areas of F-actin (cells 5 and 6).

Relationship of the nucleus to the position of the actin ring

When mutant *cdc* 11 is incubated at the restrictive temperature, cells do not divide but proceed to elongate and undergo DNA synthesis and nuclear division (Nurse *et al.*, 1976). The cells thus become multinucleate, the nuclei often becoming clustered at the centre of the cell or spaced either side of thc cell equator (*Figure 10*, cell 1). We have analysed the distribution of actin in cells held for 4 h at the restrictive temperature. By this stage, many cells are binucleate and about to enter the next round of mitosis. During the preceding interphase, F-actin is localized as clusters of dots exclusively at the cell poles (*Figure 10*, cell 1). However, when the nuclei enter the subsequent synchronous mitosis, end staining disappears to be replaced by two actin rings, presumably in the positions previously occupied by the pre-mitotic nuclei (cell 2).

Another mutant that shows a striking association of F-actin with the nucleus is the cold-sensitive, benomyl-resistant strain *ben* 4 (Roy and Fantes, 1982). At its restrictive

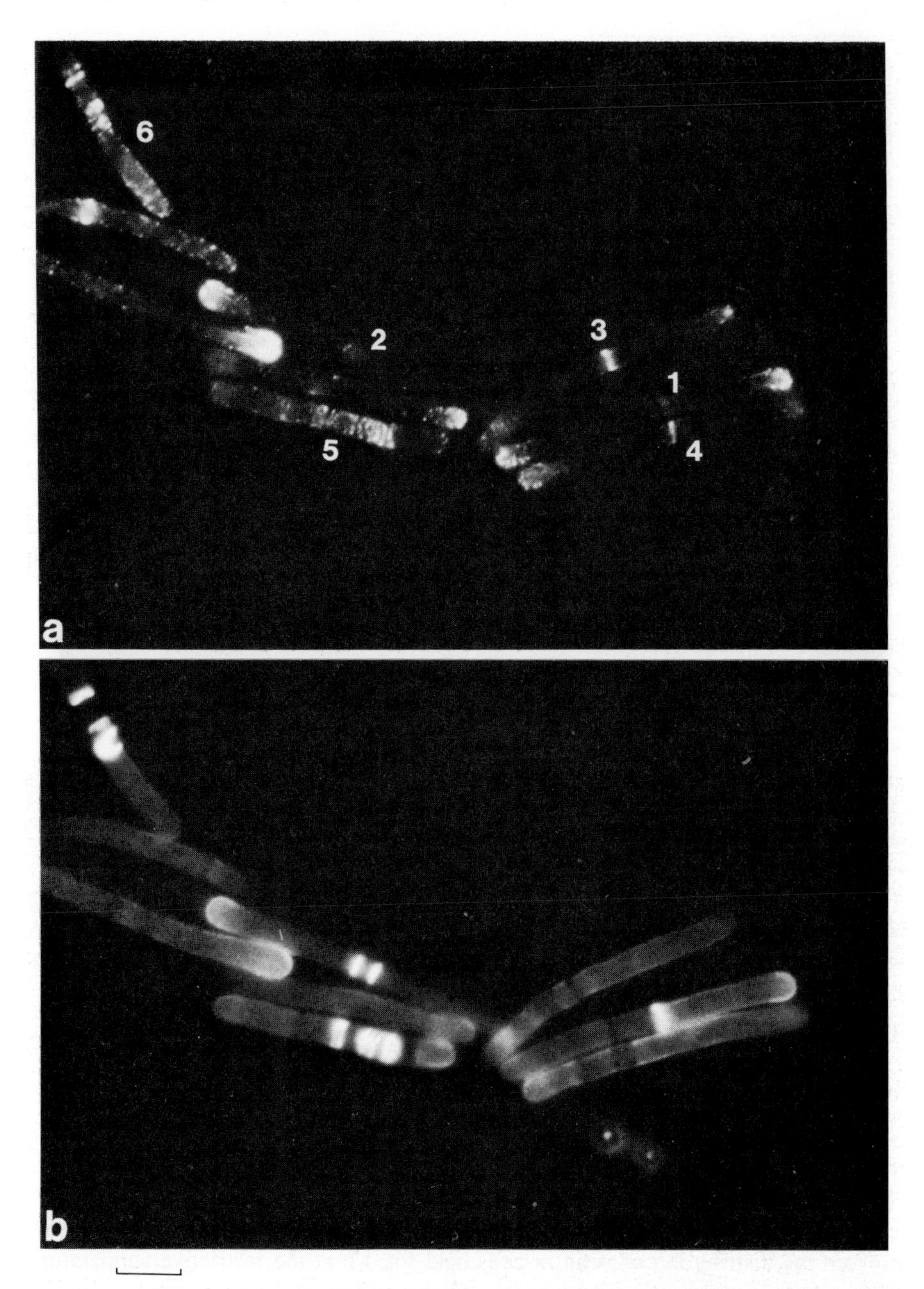

Figure 9. *cdc* 13 cells blocked at 36°C for 7 h; (**a**) stained with rhodamine-conjugated phalloidin, (**b**) stained with calcofluor. Cells show a similar pattern of actin distribution to those exhibited by wild type cells, with actin associated with polar growth or as rings preceding the deposition of the septa. The latter disperse back to the poles following septation. Scale bar = 10 μm.

temperature of 20°C, this mutant continues to grow, but fails to go through cytokinesis, and the nucleus shows substantial elongation. Under these conditions, actin is visualized as a filamentous ring which, unlike those seen in wild type cells and the *cdc* mutants,

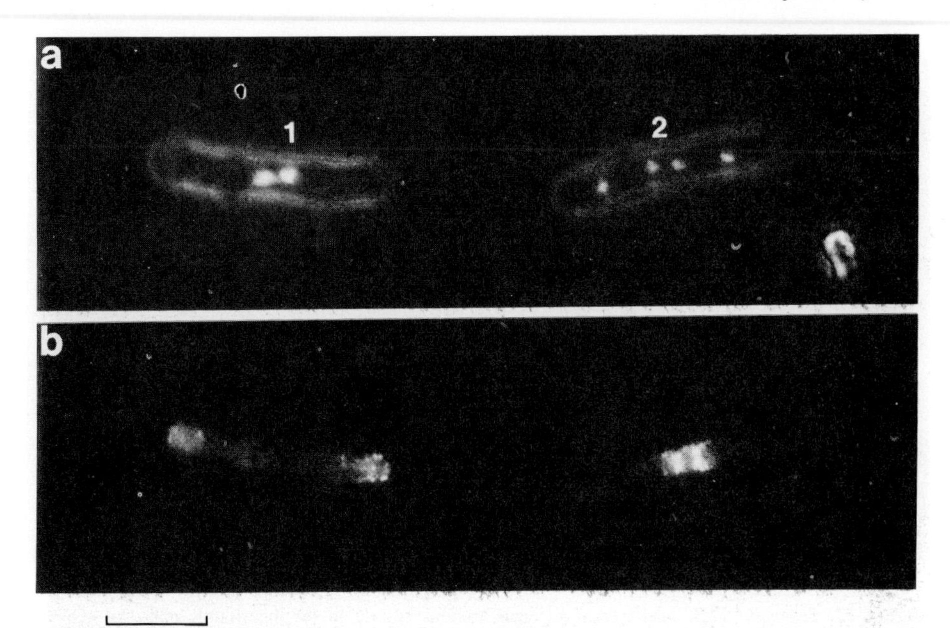

Figure 10. Temperature-arrested *cdc* 11 cells incubated at their restrictive temperature for 4 h; (**a**) DAPI and phase contrast image; (**b**) cells stained with rhodamine-conjugated phalloidin. During interphase, cells exhibit exclusively polar actin (cell 1), which reappears as two equatorial rings at the onset of the synchronous mitosis in a binucleate mother cell (cell 2). Note the difference in the intensity of DAPI staining in the pre-mitotic (cell 1) and mitotic (cell 2) nuclei. Scale bar = 10 μm.

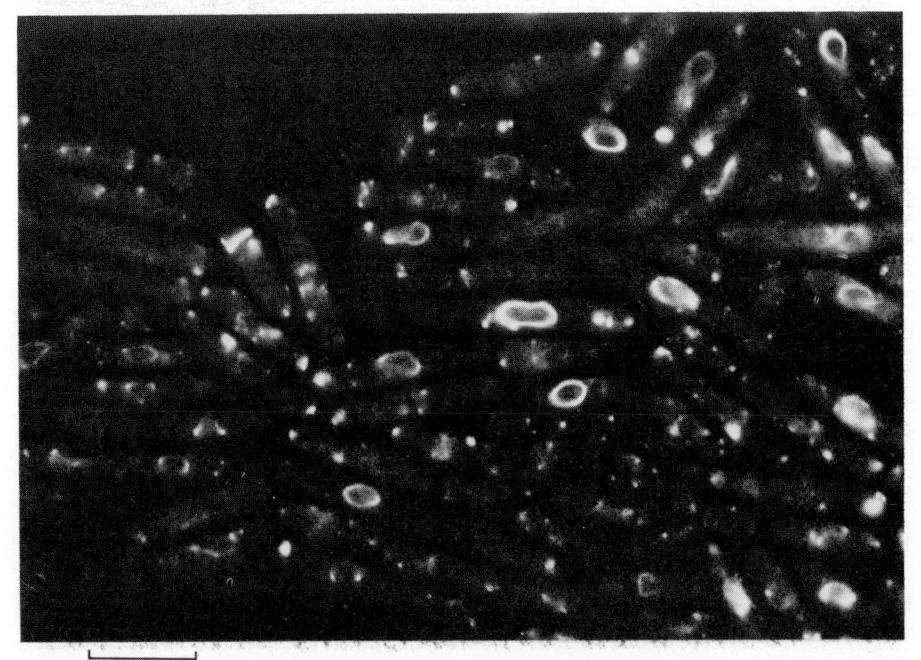

Figure 11. Rhodamine-conjugated phalloidin-stained temperature-arrested cells of the cold-sensitive mutant *ben* 4. In this mutant, actin is visualized as large bundles of filaments forming a ring around the nucleus and trailing off into the cytoplasm and also as large polar dots. Scale bar = 10 μm.

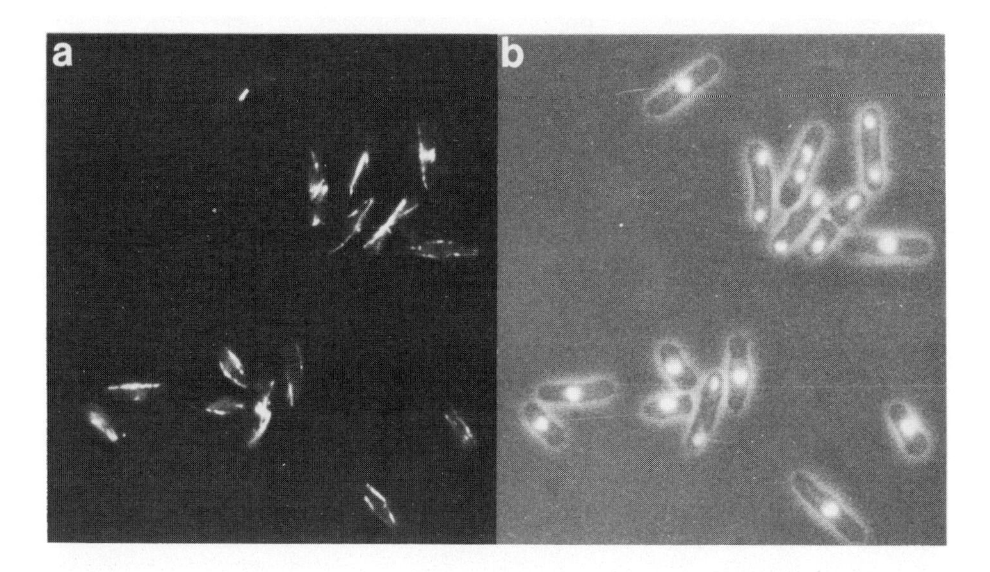

Figure 12. *S. pombe* 972h$^-$ cells from an asynchronous culture in the logarithmic phase of growth; (**a**) stained with the anti-tubulin monoclonal antibody YOL 1/34, (**b**) nuclei stained with DAPI and cell outlines visualized by phase contrast microscopy. Scale bar = 10 μm.

is oriented not around the circumference of the cell but rather in a random plane around the enlarged nucleus (*Figure 11*). Such rings can appear contorted into more than one plane, giving the appearance of a figure 8. Filaments are often seen emerging from the ring into the cytoplasm. Large F-actin dots are also seen at the cell poles, which are presumably the result of fusion or association between individual dots.

Distribution of tubulin in wild type cells

Although not excluding other possibilities, we have begun to investigate whether the changing patterns of actin distribution through the *S. pombe* cell cycle described above are associated with microtubules. *S. pombe* cells stained with the antitubulin antibody YOL 1/34 (Kilmartin *et al.*, 1982) are shown in *Figure 12*). During interphase, groups of microtubules extend between the two ends of the cells. As the cell enters mitosis, these cytoplasmic microtubules disappear, to be replaced by an intranuclear spindle. The precise details of the cell cycle rearrangements of tubulin in *S. pombe* will be presented elsewhere (I.M.Hagan and J.S.Hyams, in preparation).

Summary and speculation

Fluorescence microscopy offers some notable advantages over electron microscopy for the analysis of cytoskeletal organization, particularly in yeasts where the density of the cytoplasm is a considerable hindrance to the resolution of cytoplasmic filaments. We have used this technique to show an explicit spatial coincidence between F-actin distribution and cell wall growth in the fission yeast *S. pombe*. This relationship, first established in wild type cells (Marks and Hyams, 1985), is confirmed here by the study of a variety of morphogenetic mutants. We have shown identical patterns of actin distribution in *S. pombe* using two probes, phalloidin, which is specific for F-actin (Wieland, 1986), and a monoclonal anti-actin antibody. Since each requires quite different preparative techniques (Marks and Hyams, 1985; I.M.Hagan, and J.S.Hyams, in preparation) there seems little doubt that the patterns we describe reflect the true distribution of actin in this yeast. Indeed, our findings are so unambiguous that actin can now be used as a reliable cytoplasmic marker for the site(s) of deposition of the fission yeast cells wall. For instance, it has enabled us to resolve the long-standing controversy regarding the contribution of growth at the new end before NETO (Miyata *et al.*, 1986). Mitchison and Nurse (1985) state that in the period following cytokinesis, growth at the new end makes a minor contribution to cell extension. Accordingly we have shown that whilst actin is predominantly associated with the old end during this period a small amount of fluorescence at the new end is always detected. This is also clearly seen in mutant *cdc* 10, which arrests in G1, before the switch to bipolar growth (Aves *et al.*, 1985).

The spatial association of F-actin and cell wall deposition reported here for the fission yeast *S. pombe* has also been observed in the budding yeast *Saccharomyces cerevisiae* (Kilmartin and Adams, 1984; Adams and Pringle, 1984; Novick and Botstein, 1985), the dimorphic yeast *Candida albicans* (Anderson and Soll, 1986), and at least two filamentous basidiomycetes (Hoch and Staples, 1983a, 1985; Runberg *et al.*, 1986). The most conspicuous feature of actin organization in all of these instances is that it is primarily in the form of 'dots'. These may correspond to a class of cell wall precursor vesicles termed 'filasomes' which have been shown by electron microscopy to share a similar localization in the cortical regions of the tips of growing cells, and to be coated with filaments with the dimensions of F-actin (Hoch and Staples, 1983a,b). That the dots are real intracellular structures is suggested by the fact that they can be isolated as discrete entities (Hoch and Staples, 1983a). However, it is also possible that they are a form of 'focal contact' (King and Maciver, Chapter 7) between a cytoplasmic network of actin fibres and the plasma membrane (Adams and Pringle, 1984). This might serve to anchor the membrane to the cell wall, either as part of the wall assembly machinery or, more simply, be an intracellular scaffold imparting tensile strength to an intrinsically weak region of the cell.

A further possibility is that actin organizes and/or immobilizes ion channels in the plasma membrane at the growing cell poles (Harold and Harold, Chapter 2). Whilst an association between polar actin and endogenous electrical fields has not been demonstrated in fungi, in other tip-growing cells the correlation is clear. Most notable are experiments with germinating zygotes of the brown algae *Pelvetia* and *Fucus*. These have shown that treatment of rhizoids with the anti-microfilament drugs cytochalasin D and dihydrocytochalasin B disrupts both the normal pattern of actin localization and the endogenous electrical current (Brawley and Robinson, 1985).

Irrespective of function, any consideration of actin in fission yeast is complicated by its choreographed rearrangements through the cell division cycle, paralleling the switches in cell wall deposition. At present, we do not know whether these changes involve polymerization and depolymerization of a cytoplasmic G-actin pool or the physical redistribution of cellular components. For instance, one could envisage a situation at NETO in which a proportion of the actin at the old end (presumably associated with some type of vesicle) was transported to the new end to initiate bipolar growth.

An expectation of such a model would be that growth at the old end would decrease, because there is now less actin at this site, but the overall rate of cell elongation would remain constant as the total amount of actin is unchanged and only its distribution has altered. Analysis of living wild type cells has shown that NETO is indeed accompanied by a slowing down of growth at the old end although total growth rate increases by about 35% at this point (Mitchison and Nurse, 1985). Thus any model assuming the movement of actin-containing organelles within the cell would also have to invoke an amplification of the number of functional components at NETO. Some of these points may become clear when current experiments designed to determine the pattern of actin synthesis through the fission yeast cell cycle and the characterization of actin-binding proteins in these cells are complete. However, before leaving this speculation, it is interesting to consider how actin vesicles could be moved between the cell ends. We have shown that actin filaments appear transiently with the redistribution of actin and that cytoplasmic microtubules extend between the two poles. Both cytoskeletal elements have been shown to be involved in the directed translocation of membrane-bound organelles (Wiliamson, 1986). Further, treatment of *S. pombe* with anti-microtubule drugs induces grossly aberrant cell wall organization (Walker, 1982).

We have been unable to estimate the time required for the hypothetical movement of actin from the old to the new end. Analysis of many thousands of cells has failed to reveal obvious intermediate states and this may reflect its rapidity (or the fact that it does not occur). A few dots are always observed between the cell ends but these probably reflect a low level of non-polar wall turnover or repair (Johnson, 1965). Following cytokinesis in *S. pombe*, there is a lag in the initiation of growth at the old end (Streiblova and Wolf, 1972; Mitchison and Nurse, 1985) which might represent the time taken for actin to redistribute from the mother cell equator. However it could equally reflect a period of re-establishment of the cytoskeleton (and hence the intracellular transport machinery) following the gross rearrangements at cell division.

An equatorial ring of F-actin appears to be a ubiquitous component of the cytokinetic apparatus of eukaryotic cells (Girbardt, 1979; Marks and Hyams, 1985; Clayton and Lloyd, 1985; Rappaport, 1986). In cells with flexible boundaries, a contractile ring of actin and myosin constricts the cell surface at division. Until myosin is positively identified and localized in walled cells (Watts *et al.*, 1985), the function of the equatorial ring remains uncertain. What is clear in *S. pombe* is that the ring (a) dictates the position at which the septum will form and (b) is associated with the nucleus. The former is best demonstrated by the various *cdc* mutants which block around the time of septation (Nurse *et al.*, 1976). In mutants *cdc* 3, 4, 12, 13 and 16, which display a variety of cytokinetic defects at the restrictive temperature, aberrant septation is always accompanied by a corresponding array of actin. It is interesting that in mutants such as *cdc* 3, actin returns to the cell equator although it is not organized into a discrete ring. It is possible that such mutants may define a class of actin-binding proteins (Pollard and Cooper, 1986) and experiments to test this possibility are in progress.

The relationship of the actin ring to the nucleus is best demonstrated in the mutants *cdc* 11 and *ben* 4 (Marks *et al.*, 1986). In multinucleate *cdc* 11 cells, a ring forms in association with each nucleus at its position at the start of mitosis even when these are no longer sited at the equator. The *ben* 4^{+} gene product is apparently involved in the normal positioning of the ring since in the absence of this function the ring displaces to the surface of the nucleus. The *ben* 4 mutant is interesting in that it was isolated on the basis of its resistance to benomyl and cold-sensitivity, features typically associated with tubulin gene mutations in fungi (Morris, 1986). However, *ben* 4 maps at a site distinct from the *S. pombe* tubulin genes (P.Fantes, personal communication; see also Yanagida *et al.*, 1985). An attractive possibility is, therefore, that the *ben* 4^+ gene defines a protein that interacts with microtubules, that is, a microtubule-associated protein, which also has some association with actin. Clearly there is a relationship between microtubules and the nucleus in *S. pombe* since, in cells lacking microtubules, the nucleus is free to wander from its normal equatorial position (Hiraoka *et al.*, 1984; I.M.Hagan and J.S.Hyams, in preparation). The equatorial actin ring is not only spatially associated with the nucleus but its formation is coincident with the earliest stages of mitosis detectable by DAPI staining (Marks and Hyams, 1985). In *cdc* 13, these events are uncoupled and cells form multiple rings and septa in the absence of nuclear division (Nasmyth and Nurse, 1981). It is interesting that, although *cdc* 13 cells do not undergo cytokinesis, actin nevertheless cycles from the equator to the cell ends and back to the equator. This is shown even more clearly in *cdc* 11 mutants where it almost certainly accounts for the 'pulses' in growth between successive mitoses (Mitchison and Nurse, 1985).

In summary, we have shown that there is an intimate spatial relationship between F-actin and growth polarity and septation in the fission yeast *Schizosaccharomyces pombe*. These observations extend the study of growth in *S. pombe*, which dates back almost fifty years (Knaysi, 1940), and go some way to resolving the controversy that surrounds it (Miyata *et al.*, 1986). They also suggest ways in which similar questions might be approached in other tip-growing organisms. Finally, they provide a number of new morphological markers of progress through the fission yeast cell division cycle and hence further the usefulness of *S. pombe* for studies of cell growth and division.

References

Adams,A.E.M. and Pringle,J.R. (1984) Relationship of actin and tubulin distribution to bud growth in wild-type and morphogenetic mutant *Saccharomyces cerevisiae*. *Journal of Cell Biology*, **98**, 934–945.

Anderson,J.M. and Soll,D.R. (1986) Differences in actin localization during bud and hypha formation in the yeast *Candida albicans*. *Journal of General Microbiology*, **132**, 2035–2047.

Aves,S.J., Durkacz,B.W., Carr,A. and Nurse,P. (1985) Cloning, sequencing and transcriptional control of the *Schizosaccharomyces pombe cdc10* 'start' gene. *EMBO Journal*, **4**, 457–463.

Biely,P., Kovarik,J. and Bauer,S. (1973) Cell wall formation in yeast. An electron microscopic autoradiographic study. *Archiv für Mikrobiologie*, **94**, 365–371.

Bonatti,S., Simili,M. and Abbondandolo,A. (1972) Isolation of temperature sensitive mutants of *Schizosaccharomyces pombe*. *Journal of Bacteriology*, **109**, 484–491.

Brawley,S.H. and Robinson,K.R. (1985) Cytochalasin treatment disrupts the endogenous currents associated with cell polarization in fucoid zygotes: studies of the role of F-actin in embryogenesis. *Journal of Cell Biology*, **100**, 1173–1184.

Clayton,L. and Lloyd,C.W. (1985) Actin reorganization during the cell cycle in meristematic plant cells. *Experimental Cell Research*, **156**, 231–238.

Darken,M.A. (1961) Applications of fluorescent brighteners in biological techniques. *Science*, **133**, 1704–1705.

Girbardt,M. (1979) A microfilamentous septal belt (FSB) during induction of cytokinesis in *Trametes versicolor* (L. ex Fr). *Experimental Mycology*, **3**, 215–228.

Hiraoka,Y., Toda,T. and Yanagida,M. (1984) The NDA 3 gene of fission yeast encodes β-tubulin. A cold sensitive nda 3 mutation reversibly blocks spindle formation and chromosome movement in mitosis. *Cell*, **39**, 349–358.

Hoch,H.C. and Staples,R.C. (1983a) Visualization of actin *in situ* by rhodamine-conjugated phalloin in the fungus *Uromyces phaseoli. European Journal of Cell Biology,* **32**, 52–58.

Hoch,H.C. and Staples,R.C. (1983b) Ultrastructural organization of the non-differentiated uredospore germling of *Uromyces phaseoli* variety typica. *Mycologia,* **75**, 795–824.

Hoch,H.C. and Staples,R.C. (1985) The microtubule cytoskeleton in hyphae of *Uromyces phaseoli* germlings: its relationship to the region of nucleation and to the F-actin cytoskeleton. *Protoplasma,* **124**, 112–122.

Johnson,B.F. (1965) Autoradiographic analysis of regional cell wall growth of yeasts *Schizosaccharomyces pombe. Experimental Cell Research,* **39**, 613–624.

Johnson,B.F. (1968) Morphometric analysis of yeast cells. II. Cell size of *Schizosaccharomyces pombe* during the growth cycle. *Experimental Cell Research*, **49**, 59–68.

Johnson,B.F. and McDonald,I.J. (1983) Cell division: a separable cellular subcycle in the fission yeast *Schizosaccharomyces pombe. Journal of General Microbiology,* **129**, 3411–3419.

Kilmartin,J.V. and Adams,A.E.M. (1984) Structural rearrangements of tubulin and actin during the cell cycle of the yeast *Saccharomyces. Journal of Cell Biology*, **98**, 922–933.

Kilmartin,J.V., Wright,B. and Milstein,C. (1982) Rat monoclonal antitubulin antibodies derived by using a new nonsecreting rat cell line. *Journal of Cell Biology,* **93**, 576–582.

Knaysi,G. (1940) A photomicrographic study of the rate of growth of some yeasts and bacteria. *Journal of Bacteriology,* **40**, 247–253.

Maeda,H. and Ishida,N. (1967) Specificity of binding of hexapyranosyl sugars to polysaccharides with fluorescent brightener. *Journal of Biochemistry,* **62**, 276–278.

Marks,J., Hagan,I.M. and Hyams,J.S. (1986) Growth polarity and cytokinesis in fission yeast: the role of the cytoskeleton. *Journal of Cell Science*, Supplement **5**, 229–241.

Marks,J. and Hyams,J.S. (1985) Localization of F-actin through the cell division cycle of *Schizosaccharomyces pombe. European Journal of Cell Biology,* **39**, 27–32.

May,J.W. (1962) Sites of cell wall extension demonstrated by the use of fluorescent antibody. *Experimental Cell Research,* **27**, 170–172.

Minet,M., Nurse,P., Thuriaux,P. and Mitchison,J.M. (1979) Uncontrolled septation in a cell division cycle mutant of the fission yeast *Schizosaccharomyces pombe. Journal of Bacteriology,* **137**, 440–446.

Mitchison,J.M. (1957) The growth of single cells. I. *Schizosaccharomyces pombe. Experimental Cell Research*, **13**, 244–262.

Mitchison,J.M. (1970) Physiological and cytological methods for *Schizosaccharomyces spombe*. In *Methods in Cell Physiology.* D.M.Prescott (ed.), Academic Press, London, Vol. 4, p. 131.

Mitchison,J.M. and Nurse,P. (1985) Growth in cell length in the fission yeast *Schizosaccharomyces pombe. Journal of Cell Science,* **75**, 357–376.

Miyata,M., Miyata,H. and Johnson,B.F. (1986) Patterns of extension growth of the fission yeast *Schizosaccharomyces pombe. Canadian Journal of Microbiology,* **32**, 528–530.

Morris,N.R. (1986) The molecular genetics of microtubule proteins in fungi. *Experimental Mycology,* **10**, 77–82.

Nasmyth,K. and Nurse,P. (1981) Cell division cycle mutants altered in DNA replication and mitosis in the fission yeast *Schizosaccharomyces pombe. Molecular and General Genetics,* **182**, 119–124.

Novick,P. and Botstein,D. (1985) Phenotypic analysis of temperature sensitive yeast actin mutants. *Cell,* **40**, 405–416.

Nurse,P., Thuriaux,P. and Nasmyth,K. (1976) Genetic control of the cell division cycle of the fission yeast *Schizosaccharomyces pombe. Molecular and General Genetics,* **146**, 167–178.

Pollard,T.D. and Cooper,J.A. (1986) Actin and actin binding proteins. A critical evaluation of mechanisms and functions. *Annual Review of Biochemistry,* **55**, 987–1035.

Rappaport,R. (1986) Establishment of the mechanisms of cytokinesis in animal cells. *International Review of Cytology,* **105**, 245–281.

Roy,D. and Fantes,P.A. (1982) Benomyl resistant mutants of *Schizosaccharomyces pombe* cold sensitive for mitosis. *Current Genetics,* **6**, 195–201.

Runberg,P., Raudaskoski,M. and Virtanen,I. (1986) Cytoskeletal elements in the hyphae of the basidiomycete *Schizophyllum commune* visualized with indirect immunofluorescence and NBD-phallicidin. *European Journal of Cell Biology,* **41**, 25–32.

Streiblova,E. (1984) The yeast cell wall – a marker system for cell cycle controls. In *The Microbial Cell Cycle.* Nurse,P. and Streiblova,E. (eds), CRC Press, Boca Raton, Florida, pp. 89–107.

Streiblova,E., Malek,I. and Beran,K. (1966) Structural changes in the cell wall of *Schizosaccharomyces pombe* during cell division. *Journal of Bacteriology,* **91**, 428–435.

Streiblova,E., Hasek,J. and Jelke,E. (1984) Septum patterns in *ts* mutants of *Schizosaccharomyces pombe* defective in genes CDC 3, CDC 4, CDC 8 and CDC 12. *Journal of Cell Science,* **69**, 47–65.

Streiblova,E. and Wolf,A. (1972) Cell wall growth during the cell cycle of *Schizosaccharomyces pombe. Zeitschrift für Allgemeine Mikrobiologie,* **12**, 673–684.

Walker,G.M .(1982) Cell cycle specificity of certain antimicrotubular drugs in *Schizosaccharomyces pombe. Journal of General Microbiology,* **128**, 61–71.

Watts,F.Z., Miller,D.M. and Orr,E. (1985) Identification of myosin heavy chain in *Saccharomyces cerevisiae. Nature,* **316**, 83–85.

Wieland,Th. (1986) *Peptides of Poisonous Amanita Mushrooms*. Springer Verlag, Berlin.

Williamson,D.H. and Fennell,D.J. (1975) The use of fluorescent DNA binding agent for detecting and separating yeast mitochondrial DNA. *Methods in Cell Biology,* **12**, 335–351.

Williamson,R.E. (1986) Organelle movement along actin filaments and microtubules. *Plant Physiology,* **82**, 631–634.

Yanagida,M., Hiraoka,Y., Uemura,T., Miyake,S. and Hirano,T. (1985) Control mechanisms of chromosome movement in mitosis of fission yeast. In *Yeast Cell Biology*. Hicks,J. (ed.), Alan Liss, New York, UCLA Symposium Volume 33, pp.

INDEX